MODERN MILITARY HELICOPTERS
1990–Present

MODERN MILITARY HELICOPTERS
1990–Present
ATTACK • UTILITY • TRANSPORT • RECONNAISSANCE

BING CHANDLER

Copyright © 2025 Amber Books Ltd

Amber Books Ltd
United House
North Road
London N7 9DP
United Kingdom
www.amberbooks.co.uk
Facebook: amberbooks
YouTube: amberbooksltd
Instagram: amberbooksltd
X(Twitter): @amberbooks

All rights reserved. No part of this work may be reproduced, stored in a retrieval system, or transmitted in any form or by any means, electronic, mechanical, photocopying, recording, or otherwise, without the prior permission of the copyright holder.

ISBN: 978-1-83886-594-8

Project Editor: Michael Spilling
Design: Fredy Alexandrakis
Picture Research: Terry Forshaw

Printed and bound in Malaysia

Contents

Introduction	6
Attack Helicopters	8
Reconnaissance and Utility	36
Transports and Heavy Lifters	70
Naval Helicopters	94
GLOSSARY	123
INDEX	124
PICTURE CREDITS	128

Introduction

The first practical military helicopters were developed by Germany shortly before World War II, while on the Allied side Sikorsky had developed the R-4, which entered service in 1942. Although all these aircraft were underpowered, often barely able to lift themselves and a crew of two, they demonstrated the type's potential. Able to land in small unprepared areas, the R-4 was used to rescue injured troops and aircrew from Burma to the South Pacific.

There were also experiments with operating them from ships to conduct anti-submarine patrol. The attraction of an aircraft that could operate from a small platform on any ship big enough to carry it, offered obvious flexibility.

A UH-1N of the USMC's Marine Medium Helicopter Squadron 263 (HMM-263) drops frogmen from a low hover during a training exercise. Known as helo casting, this method allows personnel to be rapidly deployed into the water without specialised equipment.

Coming of age in Vietnam
Helicopters grew in size and complexity in the post-war period and during the 1956 Suez Crisis the Royal Navy and Royal Air Force conducted the first combat assault with rotary wing aircraft. This saw 16 Westland Whirlwind and six Bristol Sycamores transport 650 Royal Marines and 23 tonnes of stores from two aircraft carriers off the coast of Egypt to attack positions ashore. The military helicopter came of age in the Vietnam War where conditions, along with the introduction of smaller turboshaft engines, allowed new operating concepts to be developed. The UH-1 Huey became synonymous with the conflict during which it enabled troops to be rapidly moved between locations. Meanwhile the Boeing CH-47A Chinook moved tonnes of equipment and supplies to remote sites in a single lift, giving military planners options that were simply not available previously.

Vietnam also saw the emergence of the helicopter gunship. Initially

INTRODUCTION

transports had been modified with gun and rocket pods to escort troop-carrying models and provide suppressing fire during opposed landings. These were replaced by the AH-1 Cobra, which was designed specifically for the role with turret-mounted guns and tandem seating for the crew, a format that would become almost universal for attack helicopters in the coming decades.

Naval helicopters

In the maritime sphere the ability to operate from the back of a frigate or destroyer saw naval helicopters become widespread and extended the range of their parent ship's sensors. For the anti-submarine role, dipping sonars were developed to detect submarines before they were in a position to threaten friendly shipping. Radar-equipped helicopters meanwhile could find enemy shipping far over the horizon and either attack it themselves or provide target reporting.

Advanced capabilities

Modern military helicopters have further improved these capabilities with advanced sensors and weaponry. More powerful engines and advanced rotor blades have also allowed a gradual increase in their performance. However, due to the limitations of rotary wing flight the maximum speeds achievable have remained relatively static. Consequently, aircraft such as the CH-47 Chinook, UH-1 Huey and Mi-8 Hip are still in production more than 60 years after they were introduced, technology having allowed them to carry greater payloads and remain militarily relevant. As with the earliest helicopters, their ability to operate

Two Mil Mi-35M Hind attack helicopters of the Russian Air Force seen at Kubinka Air Force Base. The Mi-35M is the latest version of the 'Hind' family to enter Russian service and is fitted with updated avionics and sensors for the night attack role.

in small unprepared locations and, when necessary, remain stationary in the sky gives the military an irreplaceable capability.

ATTACK HELICOPTERS

Designed to provide overwhelming firepower, attack helicopters are highly specialized, in most cases with a limited ability to carry out any other role. Some such as the HAL Prachand and Harbin Z-19 are based on civilian designs, which make them cheap to develop and operate. At the other end of the scale, the AH-64 Apache and Mi-28N Havoc, which were developed from the ground up for the role, are armoured and have integrated sensor suites to allow attacks at night and in poor weather.

This chapter includes the following helicopters:

- AH-64 Apache
- Eurocopter Tiger
- Kamov Ka-50
- Kamov Ka-52
- Boeing AH-6
- MH-6 Little Bird
- Bell AH-1 SuperCobra
- Bell AH-1Z Viper
- Mil Mi-24
- Mil Mi-35
- Mil Mi-28N
- Changhe Z-10
- HAL Prachand
- Harbin Z-19

Left: An AH-64E Apache Guardian from the US Army's 25th Aviation Regiment conducts deck landing qualifications on the flight deck of amphibious assault ship USS *Peleliu*, 2014.

ATTACK HELICOPTERS

Boeing AH-64 Apache

Successor to the AH-1 Cobra, the AH-64 has set the standard in attack helicopters for over 40 years. Still in production, it has evolved from an analogue Cold War warrior to a networked digital fighter fit for the 21st-century battlefield.

After the initial success of the AH-1 in Vietnam the US Army launched the Advanced Aerial Fire Support System (AAFSS) programme. This would produce the AH-56A Cheyenne, a large compound rotorcraft with a pusher propeller intended to operate as a medium-level close support aircraft for gun and missile attacks. However, while the US Air Force (USAF) took issue with the Army acquiring an aircraft that infringed on its role, and the AH-56A suffered numerous technical problems, the emergence of the SA-7 Grail shoulder-launched missile rendered the proposed tactics untenable leading to the Cheyenne's cancellation.

Instead, after a re-evaluation of the requirement in 1972, the US Army launched the Advanced Attack Helicopter (AAH) programme. The request for proposals called for an aircraft able to cruise at 270km/h (145kts) with a load of eight TOW (tube-launched, optically tracked, wire-guided) missiles and an endurance of 1.9 hours. To meet the manoeuvrability requirements the helicopter needed to achieve g-limits of +3.4 to -1.5 while being strong enough to resist hits from 12.7mm (0.5in) armour piercing rounds and remain flyable after a hit from a 23mm (0.9in) anti-aircraft round. The SAM threat was to be countered by minimizing the infrared signature and the use of chaff and flare countermeasures. If this was insufficient the crew should be protected against a 48km/h (26kts) crash equivalent to a vertical impact at 12.8m/s (42ft/s) at 28km/h (15kts).

Hellfire requirement

Lockheed, Boeing, Bell, Sikorsky and Hughes all submitted bids with Bell and Hughes being selected to proceed with flying prototypes in June 1973. Both companies flew their first aircraft in September 1975, but while test and evaluation flying was ongoing, the Army added an additional requirement in early 1976, integration of the Helicopter Launched Fire-and-Forget (Hellfire) missile. This allowed targets to be engaged at ranges over 6km (3.7 miles), more than twice that of TOW, using a laser guidance system. Despite being the subject of its own extended development programme, the delay

Boeing AH-64D Longbow Apache
Although the Longbow fire control radar mounted above the main rotor is a distinctive feature of the AH-64D, it can be removed when not required. This can be to reduce weight or minimise wear on the radar itself.

AH-64E Apache
Dimensions: Length: 17.73m (58ft 2in); Rotor Diameter: 14.63m (48ft); Height: 3.87m (12ft 8in)
Weight: 10,433kg (23,000lb) maximum take-off
Powerplant: 2 × General Electric T700-GE-701D turboshaft engines, 1487kW (1994shp) each
Maximum speed: 365km/h (227mph)
Range: 1896km (1178 miles) ferry range
Service ceiling: 6100m (20,000ft)
Crew: 2
Armament: 1 × 30mm (1.18in) M230 chain gun, AGM-114 Hellfire and AGM-65 Maverick air-to-ground missiles, Hydra 70, CRV and APKWS 70mm (2.75in) rockets, AIM-92 Spike air-to-air missiles

in integrating Hellfire was considered justified given the increase in capability it would bring. In fact, the integration added five months and $49.6 million to the schedule.

On 10 December 1976 Hughes was declared the winner of the AAH

ATTACK HELICOPTERS

competition with the AH-64, the Apache name being chosen in late 1981. The aircraft has a tandem cockpit for the crew with the pilot sitting behind the co-pilot/gunner, only 60cm (2in) in front of the main rotor mast. The angular canopy comprises multiple flat panels, which limits the angles from which reflected light can give away the aircraft's position, a feature first seen on the AH-1S. With US doctrine now emphasizing the ability to fight at night the AH-64 featured multiple electro-optical sensors. Below the nose is the target acquisition and designation sight (TADS) containing an infrared camera and a monochrome daylight camera, for use by the co-pilot/gunner. Above the nose is the pilot's night vision system (PNVS), which contains an infrared camera slaved to their head with the image projected onto a helmet-mounted display.

Avionics for these systems are in the distinctive cheek fairings, which have grown with subsequent models as additional capabilities have been added. Since 2005 the Arrowhead second generation TADS/PNVS has been rolled out to the AH-64D fleet and feature modern higher resolution sensors. Below the fuselage is the 30mm (1.18in) chain gun, which can also be slaved to the crews' heads while in the event of a crash the weapon collapses into a space between the pilots' seats. Stub wings each carry two pylons for external stores; wingtip carriage of Sidewinder

McDonnell Douglas AH-64 Apache

Israel has operated the AH-64A since 1990 where it is known as the Peten, or Python, and since 2005 has supplemented them with D models, known as Saraf, or Serpent. These are almost always seen with the yellow V-shaped identification marking.

11

or Stinger missiles was proposed to the US Army and although not taken up by them have been adopted by some foreign operators. The Apache was one of the first aircraft to feature an X configuration tail rotor in which the four blades are not evenly spaced. This reduces the noise of the rotor system by increasing the spacing between the main and tail rotor blades when they pass reducing the interaction between their tip vortices.

To allow for rapid deployment the AH-64 was designed to be air transportable. In transport configuration, with rotors, radar and stub wings removed, three AH-64 can be carried in a C-17 along with two cargo pallets and 50 personnel. Without removing the rotors and wings, two can still be carried along with four pallets and 38 personnel. As part of the Cold War requirement to reinforce Western Europe, the Apache is also capable of self-deploying across the Atlantic in 1480-km (800-nm) legs via Canada, Greenland, Iceland and Scotland.

Action in Panama
In 1989 the Apache, along with the F-117 stealth fighter, saw combat for the first time during the US invasion of Panama to depose General Noriega. This would also see the first operational use of Hellfire with aircraft of 'B' Company, 1st Battalion, 82nd Airborne Division firing two through the windows of Noriega's headquarters, demonstrating its precision strike capability.

The Apache truly came of age in the 1991 Gulf War where it can justifiably claim to have fired the first shots. Just after midnight on 17 January, eight AH-64A and two MH-53J, crossed the border into southern Iraq. Using TADS/PNVS and night vision goggles the Apaches kept formation on the MH-53J whose superior navigation suite was used to guide the whole force. To allow the Apaches to update their Doppler navigation systems the MH-53J dropped chemical lights, or glow sticks, at pre-arranged positions. Known as Task Force Normandy, the aircraft's mission was to destroy two Iraqi radar sites and allow strike aircraft into western Iraq undetected.

To make the 1667km (1036 miles) round trip without a refuelling stop the Apaches carried a single 870 litre (191 gallon) external tank on the right inner pylon along with eight Hellfire and 19 Hydra rockets. At 02:38 having split into their two ship attack formations, the Apaches opened fire on their targets, which had been marked with their lasers. Although initially attacking with Hellfires from 6km (3.7 miles), some aircraft ended the attack as close to their targets as 800m (0.4 miles) to use their chain guns. After four and a half minutes the attack was over, and the radar sites were destroyed.

While the AH-64A proved itself in the 1991 Gulf War where 277 saw action destroying over 500 tanks, armoured personnel carriers, and other vehicles there were shortcomings. The navigation system as seen on the opening night of the conflict was limited. At the same time the cockpit featured 1200 switches and an array of analogue gauges pushing the pilot's workload to the limit in the low-level environment for which it was intended. To address these issues and provide additional capabilities, the AH-64D model was developed. This gives the aircraft a new MIL-STD 1553B data bus, processors, and power system. The cockpits are now centred on two multifunction displays, the switch count has been reduced to around 200, while navigation is handled by an AN/ASN-157 Doppler system and dual embedded GPS. All AH-64D can mount an AN/APG-78 Longbow fire control radar

Westland WAH-64D Apache
As well as more powerful RTM engines the Westland-built Apaches featured blade folding for shipboard use and UK communications equipment. The UK's aircraft have all now been rebuilt as AH-64E.

ATTACK HELICOPTERS

Boeing Apache AH-64E
Boeing combined components from the UK's WAH-64D with new fuselages to produce 50 AH-64E for the British Army. The last of these was delivered in March 2025, the aircraft all using new serials in the ZM700-749 range.

Boeing AH-64D Apache
Kuwait received its first of sixteen AH-64D in 2005, in 2020 a contract was signed for the remanufacture of these to E models in addition to eight new build aircraft.

AH-64E Apache
Dimensions: Length: 17.73m (58ft 2in); Rotor Diameter: 14.63m (48ft); Height: 3.87m (12ft 8in)
Weight: 10,433kg (23,000lb) maximum take-off
Powerplant: 2 × General Electric T700-GE-701D turboshaft engines, 1487kW (1994shp) each
Maximum speed: 365km/h (227mph)
Range: 1896km (1178 miles) ferry range
Service ceiling: 6100m (20,000ft)
Crew: 2
Armament: 1 × 30mm (1.18in) M230 chain gun, AGM-114 Hellfire and AGM-65 Maverick air-to-ground missiles, Hydra 70, CRV and APKWS 70mm (2.75in) rockets, AIM-92 Spike air-to-air missiles

above the main rotor, this has a range of 8km (4.9 miles) and can classify and display 156 targets simultaneously while the aircraft remains behind cover. More importantly, it can be used to guide the radar-guided AGM-114L variant of the Hellfire allowing all 16 missiles to be fired in rapid succession. This is a significant advantage over the A model, which could only engage two targets at the same time and required visual line of sight to the target exposing it to enemy fire.

Further combat action
AH-64D were produced both as new builds and by re-manufacture of A models with the first examples entering service with the US Army in 1997. It was heavily involved in operations in Afghanistan from 2001 and Operation Iraqi Freedom from 2003. In both these theatres the Longbow radar was regularly removed to save weight due to the absence of enemy armoured vehicles. D Models from the British Army and Royal Netherlands Air Force also took part in these operations. The British AugustaWestland-built AH-64D were unique in being powered by the 1600kW RTM322 engine giving commonality with the AW101 Merlin, however, its AH-64E will use the same T700 engines as other operators. Foreign operators also include Israel whose aircraft have been used in an air-to-air role, Saudi Arabia, the Indian Air Force and Army, United Arab Emirates, Singapore, South Korea, Japan and Taiwan.

ATTACK HELICOPTERS

In 2012 the AH-64D Block III was redesignated the AH-64E Guardian due to the upgrade in overall capability. The fire control radar's range has doubled, Link 16 has been fitted to allow better sharing of information, and the ability to control UAVs via an L-3 Communications datalink added. The gearbox has been upgraded to handle the greater power available from the T700-GE-701D engines, which combined with new composite rotors provides an increase in cruise speed, rate of climb, and payload. To assist with the aircrew's decision-making a Cognitive Decision Aid System (CDAS) has been fitted to help with processing the data presented to them and reduce their workload. With support for the AH-64D ending almost all operators are either upgrading their aircraft to Guardian standard or purchasing new build examples.

Wings
In addition to the two underwing hardpoints the wingtips can also mount weapons such as AIM-92 Stinger, ECM equipment, or the datalink for Spike NLOS missiles.

ATTACK HELICOPTERS

Rotor Blades
Designed to absorb a hit from a 23mm round the Apache's main rotor blades have been upgraded for the AH-64E. The new composite blades improve the cruise speed, climb performance, and payload.

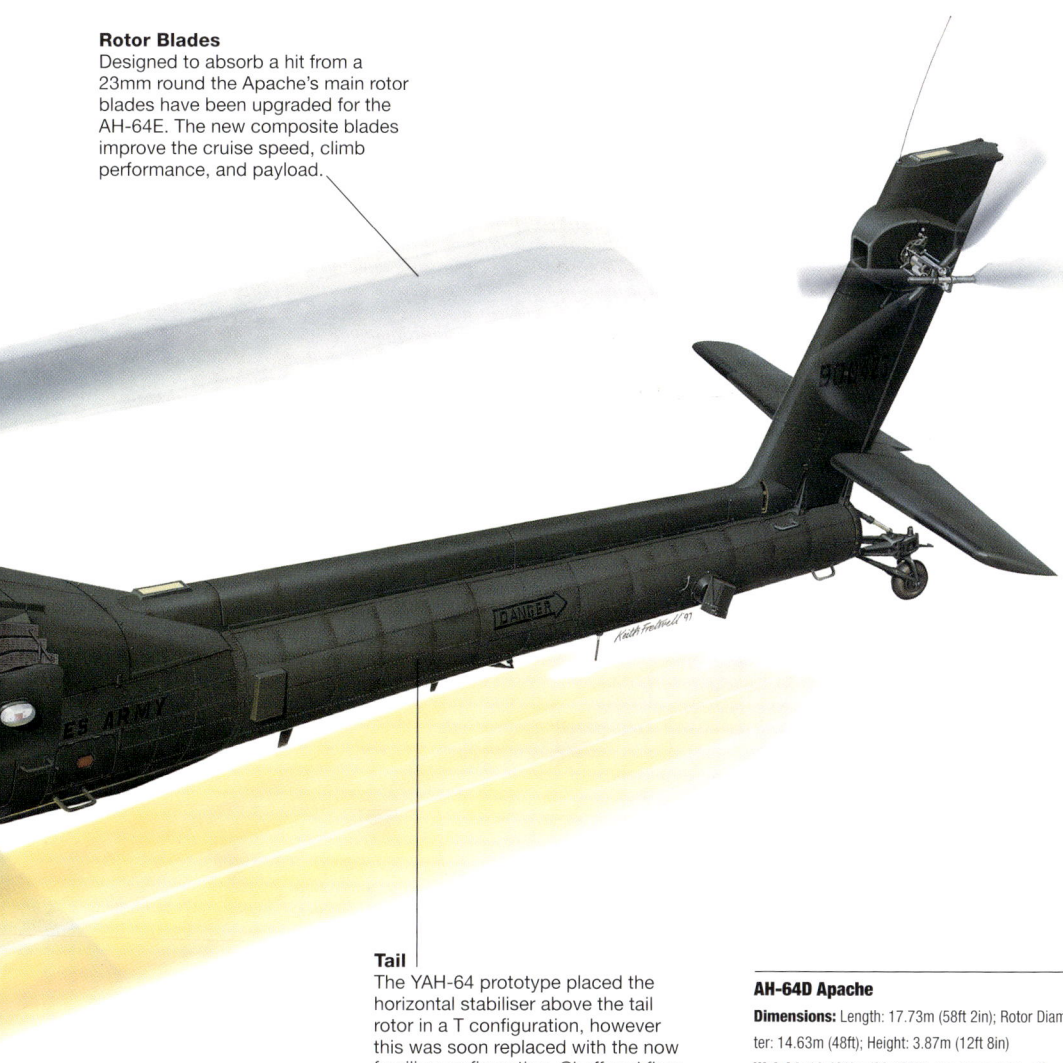

Tail
The YAH-64 prototype placed the horizontal stabiliser above the tail rotor in a T configuration, however this was soon replaced with the now familiar configuration. Chaff and flare dispensers are mounted on the sides of the tail boom just forward of the vertical stabiliser.

AH-64D Apache
Dimensions: Length: 17.73m (58ft 2in); Rotor Diameter: 14.63m (48ft); Height: 3.87m (12ft 8in)
Weight: 10,433kg (23,000lb) maximum take-off
Powerplant: 2 × General Electric T700-GE-701D turboshaft engines, 1487kW (1994shp) each
Maximum speed: 365km/h (227mph)
Range: 1896km (1178 miles) ferry range
Service ceiling: 6100m (20,000ft)
Crew: 2
Armament: 1 × 30mm (1.18in) M230 chain gun, AGM-114 Hellfire and AGM-65 Maverick air-to-ground missiles, Hydra 70, CRV and APKWS 70mm (2.75in) rockets, AIM-92 Spike air-to-air missiles

AH-64D Apache
This is the fourth development AH-64D Longbow Apache, which first flew on the 4th of October 1993 and was involved in trials and testing at Mesa, Arizona.

ATTACK HELICOPTERS

Eurocopter Tiger

The Tiger's development dates back to 1984, when France and West Germany issued a joint requirement for 427 attack helicopters. However, programme difficulties and the end of the Cold War saw the first production orders only placed in 1999, with deliveries beginning in 2005.

Initially developed by Eurocopter, the Tiger is now produced by Airbus Helicopters who designate it the EC665. Built primarily from composite materials, the Tiger is designed for battlefield survivability with a degree of radar, infrared and acoustic signature reduction. It is also able to withstand hits from 23mm (0.9in) rounds and has extensive system redundancy, while in the worst case the crew are protected by the crashworthy design. Contrary to the 1995 James Bond film GoldenEye, however, it does not feature an escape pod.

Germany's Tiger models are equipped with a mast-mounted site while unusually for an attack helicopter they do not have an integrated gun. Instead, they can carry a 20mm (0.78in) cannon pod under each stub wing. Other operators mount the electro-optical site just aft of the canopy in front of the rotor mast while carrying a 30mm (1.18in) GIAT cannon under the nose. Weapons options include guided missiles – France and Australia opting for Hellfire, Spain Spike-ER and Germany PARS 3 LR – and rockets, again the exact type varying by country.

Tigers have been used operationally in Libya, where they flew from helicopter carriers offshore, Afghanistan and Mali. Not all customers have been happy with the performance of their aircraft however. Australia notably intends to replace theirs with AH-64E Apaches from 2026, only 10 years after the Tiger achieved full operational capability with the Australian Army.

Eurocopter Tiger
F-ZWWW was the prototype Tiger and first flew on the 27th of April 1991. Used for development and sales work, it is now preserved at Le Luc in the South of France.

Eurocopter Tiger
Dimensions: Length: 14.08m (46ft 2in); Rotor Diameter: 13m (42ft 8in); Height: 3.83m (12ft 7in)
Weight: 6000kg (13,228lb) maximum take-off
Powerplant: 2 × MTR390 turboshaft engines, 972kW (1303shp) each
Maximum speed: 315km/h (196 mph)
Range: 800km (500 miles)
Service ceiling: 4000m (13,000ft)
Crew: 2
Armament: 1 × 30mm (1.18in) GIAT 30 cannon, AGM-114 Hellfire and Spike-ER air-to-ground missiles, SNEB and FZ-225 rockets, and Mistral or Stinger air-to-air missiles

ATTACK HELICOPTERS

Kamov Ka-50 / Ka-52

In response to a 1976 requirement for a Soviet attack helicopter, Kamov proposed an unusual design with a co-axial rotor system and a crew of one.

The first aircraft flew in June 1982 and testing proceeded well, however, a crash in 1985 due to blade clashing led to a redesign of the rotor system. This increased the spacing between the two rotor discs and added a feedback feature in the flight controls to increase the effort needed to move them when the blades are in danger of clashing. While improvements in the night vision system and pilot interface were recommended the aircraft was selected to proceed to production status, which began in 1990 after the completion of development flying.

Design features

The Ka-50, NATO reporting name Hokum, is a compact helicopter barely longer than its own rotors. The co-axial design allows all the power from the two engines to be used for lift, conventional helicopters using 10–20% of their power to drive the tail rotor depending on the flight regime. The engines themselves are placed on either side of the fuselage to avoid both being damaged by a single hit.

The cockpit is armoured as is the compartment containing the hydraulic systems, while the control rods are as far as possible routed inside the protection of the cockpit armour. The lack of tail rotor meanwhile eliminates a weak point of most helicopters, although unlike conventional helicopters a rudder is fitted for yaw control at high speed. If all this is insufficient to prevent the loss of an aircraft, the Ka-50, unusually for a helicopter, features an ejection seat. On activation explosive charges detach the rotor blades before firing the seat out of the cockpit.

A Shipunov 2A42 30mm (1.18in) cannon is mounted on the right-hand side of the fuselage. Unlike most attack helicopters aiming requires pointing the helicopter at the target, which is aided by the high yaw rate of the co-axial design. The cannon itself can traverse 15° in azimuth and +15°/-30° in elevation for precise targeting once the aircraft is facing in the right direction. The broad wings each carry two pylons for missiles, rockets or bombs, and SA-18 Grouse

Kamov Ka-52E 'Nile Crocodile'
6617 is operated by the 549th Air Wing of the Egyptian Air Force, based at Wadi al Jandali Air Base. Egypt has 46 Ka-52E aircraft and is currently the only export operator of the type.

Ka-52
Dimensions: Length: 15.96m (52ft 4in); Rotor diameter: 14.5m (47ft 7in); Height: 4.93m (16ft 2in)
Weight: 11,300kg (24,912lb) maximum take-off
Powerplant: 2 x Klimov VK-2500 turboshaft engines, 1800kw (2400shp) each
Maximum speed: 350km/h (217mph)
Range: 455km (283 miles)
Service ceiling: 5000m (16,400ft)
Crew: 2
Armament: 1 x 30 mm (1.18in) Shipunov 2A42 cannon, up to 2000kg (4400lb) of ordnance including AT-16 anti-tank missiles, LMUR air-to-surface missiles, SA-18 Grouse air-to-air missiles, S-8 and S-13 rockets, and 250kg (550lb) or 500kg (1100lb) bombs

ATTACK HELICOPTERS

air-to-air missiles can be carried under the wingtip ECM pods. Combat footage from Ukraine has shown Ka-52 wings vibrating significantly, although it is not clear how much of an impact this has on the attached weapons' performance.

Halt in production

While work on the initial batch of production helicopters began in 1990, the dissolution of the Soviet Union the following year virtually halted the programme. By 2000 only around a dozen airframes had been produced and attempts at international sales proved unsuccessful. At the same time Kamov had been working on a two-seat version, the Ka-52 'Alligator' or Hokum B. This mounted additional reconnaissance systems that had previously been intended for a scout helicopter. An Arbalet radar is carried in a broad nose radome while an under-nose turret houses a TV camera, thermal imager and laser designator.

The two crew sit side-by-side on ejection seats in a widened forward fuselage and both have flying controls. The first aircraft was built by modifying a production Ka-50, essentially replacing the entire forward fuselage. The cockpit features colour multifunction displays for flight and mission information with a set of centrally mounted back-up instruments above them. While just under 200 Ka-52 have been delivered to date, less than 20 Ka-50 were produced.

Ka-50s saw limited use in the Second Chechen War in 2000–2001 where it performed successfully in the mountainous terrain. The first combat use of the Ka-52 was in the Syrian Civil War where they supported Russian special forces and escorted search and rescue helicopters.

Performance in Ukraine

The invasion of Ukraine by Russia has seen extensive use of the Ka-52 by Russian forces; however, this has also led to significant losses. While there have been losses across all Russia's aviation forces, the Ka-52 has been one of the hardest hit, by various estimates losing 40–60 aircraft, or

Kamov Ka-52
Yellow 22 was operated by the Russian Air Force. However, during the assault of Hostomel airport during the invasion of Ukraine it was shot down on the 24th of February 2022.

Ka-52
Dimensions: Length: 15.96m (52ft 4in); Rotor diameter: 14.5m (47ft 7in); Height: 4.93m (16ft 2in)
Weight: 11,300kg (24,912lb) maximum take-off
Powerplant: 2 x Klimov VK-2500 turboshaft engines, 1800kw (2400shp) each
Maximum speed: 350km/h (217mph)
Range: 455km (283 miles)
Service ceiling: 5000m (16,400ft)
Crew: 2
Armament: 1 x 30 mm (1.18in) Shipunov 2A42 cannon, up to 2000kg (4400lb) of ordnance including AT-16 anti-tank missiles, LMUR air-to-surface missiles, SA-18 Grouse air-to-air missiles, S-8 and S-13 rockets, and 250kg (550lb) or 500kg (1100lb) bombs

ATTACK HELICOPTERS

Kamov Ka-52
As seen in Autumn 2022, this anonymous Ka-52 took part in the Russian invasion of Ukraine. This example has all markings painted over, but carries a white "V" recognition symbol on the engine nacelles. It may be from the 39th Guards Independent Helicopter Division, based in Dzhankoi on the Crimean Peninsula.

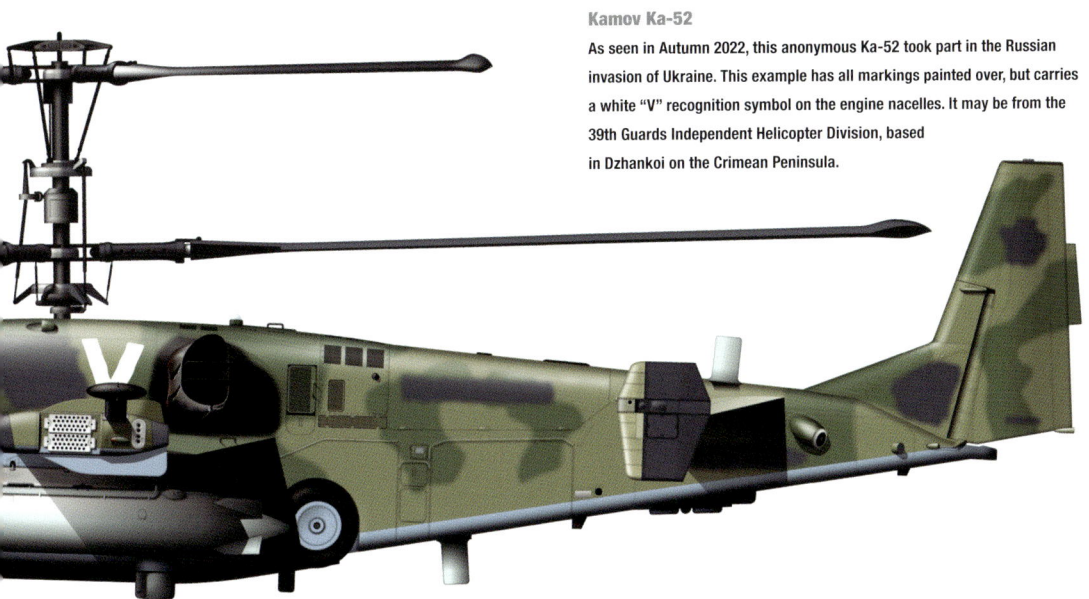

around 40% of the operational fleet of approximately 130. This compares to around 14 losses in total for the Mi-28N, which operates in a similar role and with a comparable fleet size. The high loss rate is at least in part likely due to the vulnerability of the rotor system. While the blades are designed to take damage, this can change their aerodynamics allowing them to flap out of the plane of rotation by up to 1.5m (5ft). This brings them closer to the other rotor disc making a blade clash under heavy manoeuvring more likely.

A number of foreign sales attempts have been made by Kamov, including to South Korea, and a bid for a Turkish requirement led to a completely redesigned forward fuselage with tandem seating. However, apart from Russia only Egypt has officially purchased the type, the Ka-52 'Nile Crocodile' featuring French avionics and additional cooling.

A pair of Kamov Ka-52 attack helicopters seen during rehearsals for a Victory Day parade at Kubinka Air Force Base. The large under nose sensor pod is clearly visible, although not an ideal location the co-axial rotor system does not allow for a mast mounted system.

ATTACK HELICOPTERS

AH-6/MH-6 Little Bird

The AH-6 can trace its origin back to the Hughes OH-6 Cayuse and was developed to meet a 1960 US Army Specification for a light observation helicopter.

First flying in 1963, it entered service three years later and soon saw action in Vietnam. Here it was regularly teamed with AH-1 Cobra gunship helicopters in hunter-killer pairs. The OH-6 operated at low level to flush out enemy targets, which allowed the Cobra to engage them with rockets, grenades or M134 miniguns.

The OH-6 has a distinctive teardrop-shaped fuselage that has been retained through successive generations of the basic airframe. Crash survivability was prioritized with a safety cage protecting the cabin and crew while the tail, engine and rotors were designed to break off in the event of an impact absorbing energy. These features were extensively tested in Vietnam where crews regularly survived to fight another day even after their aircraft were written off. Perhaps more remarkably, in April 1966 a modified OH-6 set the world record for long distance flight in a rotorcraft flying non-stop from Culver City, California to Ormand Beach, Florida. With over a tonne of fuel onboard the aircraft covered the 3561.55km (2213.1 miles) in 15 hours and eight minutes a record that stands to this day.

Development of MH-6

The MH-6 was developed from the OH-6 in response to the failed Iranian hostage rescue attempt in April 1980. In planning a second rescue a need was identified for a helicopter that could land in confined areas to deploy troops and that could be readily transported in US Air Force (USAF) airlifters. At the same time an armed variant was developed that would become the AH-6. Although the rescue attempt was called off after Iran released the hostages it was decided to retain the MH-6/AH-6 and the 160th Special Operation Aviation Regiment that had been formed to operate them alongside Black Hawks and Chinooks.

To transport troops the MH-6 is equipped with benches on either side of the fuselage. These can carry up to three troops each who can be deployed via fast-rope if it is not possible to find a suitable landing site. The gunship

Boeing MH-6M

Boeing MH-6M 85-25346 from the 160th Special Operations Aviation Regiment headquartered at Fort Campbell, Kentucky. This example is equipped with the Fast Rope Insertion Extraction System (FRIES) to deploy troops from the hover.

MH-6/AH-6

Dimensions: Fuselage Length: 7.5m (24ft 7in); Rotor diameter: 8.35m (27ft 5in); Height: 2.67m (8ft 9in)
Weight: 1406kg (3100lb) maximum take-off
Powerplant: 1 x Allison T63-A-5A turboshaft engines, 317kw (425shp)
Maximum speed: 282km/h (175mph)
Range: 430km (267 miles)
Service ceiling: 5700m (18,700ft)
Crew: 2
Armament: 2 x 12.7mm (0.5in) GAU-19 heavy machine gun or 2 x 7.62mm (0.3in) M134 miniguns, 70mm (2.75in) Hydra 70 rockets, AGM-114 Hellfire, FIM-92 Stinger missiles

ATTACK HELICOPTERS

configured AH-6 utilize a stub wing running through the cabin doors with two hardpoints on either side. These can carry a range of stores including rockets, missiles and gun pods for direct attacks and close air support.

Little Birds

The first operational use of the MH and AH-6, nicknamed Little Birds by the 160th, was during the 1983 invasion of Grenada prior to the public acknowledgement of their existence. Deploying to the island in four C-130 Hercules, the helicopters were used in an unsuccessful attack on the Grenadian military headquarters, medical evacuations, and search and rescue operations. Around the same time MH-6s were being used to support clandestine missions in Nicaragua with the unit's personnel wearing civilian clothes to mask US government activity. More overt activity took place during the late 1980s when Little Birds were used during the Iran–Iraq war to repel Iranian attacks on oil tankers. Operating from ships and converted barges MH-6 would scout for targets before directing the AH-6 to attack with miniguns and 70mm (2.75in) rockets.

During the invasion of Panama in December 1989 the 160th's Little Birds were used extensively and were transported to Howard AFB in USAF C-5 Galaxies. The helicopters were used to insert special operations personnel in advance of the main force before conducting attacks on the Panamanian Defence Force headquarters. As in Grenada the aircraft were also used for SAR and medevac missions as well as close air support to special forces. The aircraft were also heavily involved in the Iraq War from 2003 to 2011 supporting multiple special force operations by the USA and allied forces.

Recent developments

While initially based on the OH-6, the current MH-6M and AH-6M are based on the MD 530F, which is itself a commercial development of the original aircraft. This features a six-bladed main rotor, in contrast to the 4- or 5-bladed systems on earlier aircraft, with a four-bladed tail rotor. The tail now features a T-configured horizontal and vertical stabilizer, in contrast to the Y-shaped layout of the OH-6A. While still using the Allison T63 turboshaft this now delivers up to 317kW (425shp), a marked increase on the 188kW (252shp) of the original aircraft. This also features a full authority digital

Boeing AH-6i (AH-6SA)
Boeing AH-6i 61007 belonging to the 1st Aviation Brigade of the Saudi Arabian National Guard (SANG) based at Khashm Al An Airfield, near Ryiadh, Saudi Arabia. Saudi Arabia ordered 24 of the aircraft in August 2014.

engine control (FADEC) system to reduce pilot workload while an inlet barrier filter prevents dust and sand entering the intake and degrading performance.

Other enhancements over the original aircraft include a night vision compatible cockpit with multifunction LCD displays, control units and a cockpit management system. An AN/ZSQ-3 electro-optical turret is mounted under the nose with the imagery displayed in cockpit on the multifunction displays. The AH-6M's turret also mounts a laser rangefinder and target designator allowing AGM-114 Hellfire missiles to be used when required. The latest developments include the MH-6X Unmanned Little Bird demonstrator and AH-6I, which leverages technology developed for the Apache Block III upgrade programme and has to date been bought by Jordan, Saudi Arabia and Thailand.

ATTACK HELICOPTERS

Bell AH-1 Super Cobra

Originally developed as a helicopter gunship for the US Army, the US Marine Corps (USMC) were keen to acquire their own version with a twin-engines for operations over the sea and a 20mm (0.78in) cannon in place of the 7.62mm (0.3in) miniguns for greater hitting power.

A US Marine Corps Reserve AH-1W SuperCobra helicopter with Marine Light Attack Helicopter Squadron 773 takes off for a training mission on Joint Base McGuire-Dix-Lakehurst, N.J., October 2019.

The USMC's requirements were accepted and the AH-1J was ordered in early 1968 with the first flight occurring the following year. Entering service in 1971 the Sea Cobra saw limited operational use in the closing stages of the Vietnam War. Power was provided by a Pratt & Whitney PT6T twin-pack, this used two 1140kW (1530shp) PT6 gas turbines to drive a combining gearbox the output from which went into the main rotor gearbox. This minimized the modifications to the helicopter's transmission system while providing redundancy in the event of an engine failure and the same system was used on the UH-1N twin Huey. The AH-1J's turret meanwhile was equipped with a three-barrel 20mm (0.78in) M197 Gatling cannon to give the increased firepower the Marines required.

The AH-1J was soon followed by the AH-1T with an improved transmission system and longer fuselage and tail boom. TOW missiles were also fully integrated giving it a more effective stand-off attack capability. First flying in 1976 the AH-1T replaced the J model and saw combat during the 1983 invasion of Grenada, where they were involved in the capture of Pearls Airport and strikes on Fort Rupert and Fort Patrick. The same year saw them in action over Beirut and Lebanon in support of the peacekeeping force intervening in the civil war. In the late 1980s AH-1T also took part in the Tanker War in the Persian Gulf defending shipping from attacks by Iranian surface vessels, on one occasion sinking three patrol boats.

AH-1W Super Cobra

The early 1980s also saw the USMC looking for a replacement attack helicopter, with Congress blocking a 'marinized' AH-64 Apache on cost

ATTACK HELICOPTERS

Bell AH-1W Super Cobra
For Operation Desert Storm (1991), Super Cobras were repainted in a variety of schemes, such as this one, which replaces the blacks and greens selected for the European theatre with a sand colour more suited to the Middle East.

Bell AH-1W Supercobra
Dimensions: Fuselage Length: 14m (45ft 11in); Rotor diameter: 13.89m (45ft 7in); Height: 4.19m (13ft 9in)
Weight: 6690kg (14,750lb) maximum take-off
Powerplant: 2 x General Electric T700-401 turboshaft engines, 1260kw (1690shp)
Maximum speed: 350km/h (220mph)
Range: 587km (365 miles)
Service ceiling: 3700m (12,200ft)
Crew: 2
Armament: 1 x 20mm(0.78in) M197 Gatling cannon or 2 x 7.62mm (0.3in) M134 miniguns, 70mm (2.75in) Hydra 70 and 127mm (5in) Zuin rockets, AGM-114 Hellfire, AIM-9 Sidewinder missiles

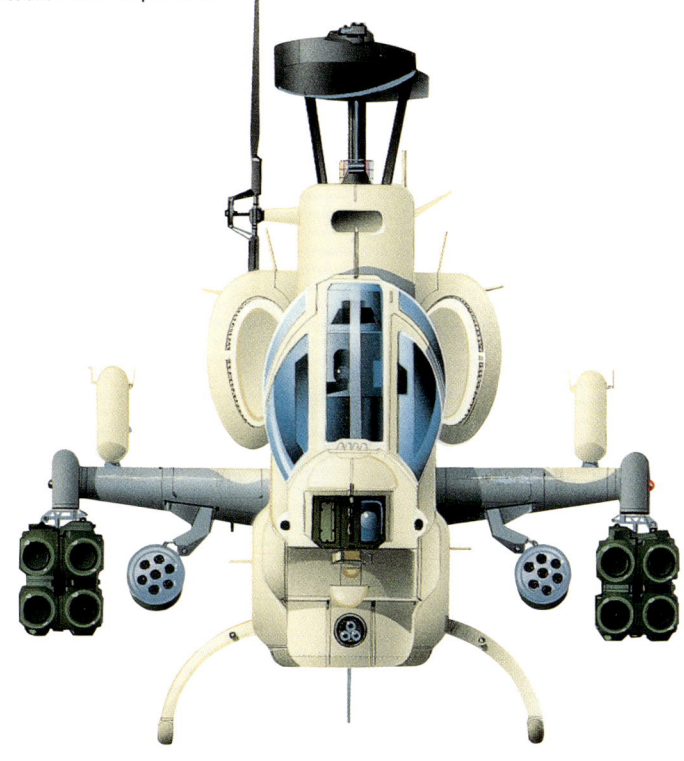

grounds a further upgrade of the AH-1 would be approved. This led to the AH-1W Super Cobra powered by 1260kW (1690shp) GE T700 engines it was also the first version able to carry AGM-114 Hellfire laser-guided missiles. The Super Cobra's maximum take-off weight is 6690kg (14,750lb) just over two tonnes more than the AH-1J's 4536kg (6610lb), which allow a greater quantity of weaponry to be carried although range and speed remain broadly similar. As well as 179 new build aircraft there was sufficient commonality with the AH-1T that 43 aircraft were upgraded to the new standard.

The AH-1W would be extensively used during the first Gulf War when 48 AH-1W and a number of AH-1T were deployed with around half being based ashore and the rest onboard assault ships. These flew a total of 1273 sorties during which the AH-1W accounted for 97 tanks and 48 armoured personnel carriers. None of the Super Cobras were lost in combat although at least one was lost in an accident. They also saw use during the US invasion of Haiti in 1994 and peacekeeping operations in the

ATTACK HELICOPTERS

Former Yugoslavia through much of the decade.

Upgrades post-Gulf War
Experience in the Gulf Wa led to upgrades to the Super Cobra to improve its capability. A night targeting system (NTS) added visual and infrared sensors, an upgraded laser rangefinder and automatic target tracking to the M65 sight mounted under the nose. A multifunction display was also added to the front cockpit for the gunner. This was rolled out to the AH-1W fleet during the mid-1990s and included a canopy cockpit modification (CCM) to replace the entire canopy, nose section and front instrument panel. A tactical navigation system was integrated as part of the upgrade, which added the AN/ASN-163 GPS/INS. Self-defence was improved with the incorporation of the AN/AAR-47 missile warning system and AN/ALE-39 chaff and flare dispenser. The former detects the heat signature of a missile being launched, alerting the crew, while the latter can automatically dispense countermeasures.

These improved Super Cobras returned to Iraq in 2003, this time 54 AH-1W were deployed and involved in heavy fighting. Despite this only two were lost to ground fire throughout the whole campaign and the aircraft demonstrated its ability to withstand

A US Marine AH-1 Super Cobra conducts live-fire training during exercise Scorpion Fire in the Chocolate Mountain Aerial Gunnery Ranges, California, 2014.

significant damage. One aircraft took a 23mm (0.9in) round to its main rotor blade but remained on station for a further four hours before returning to its forward operating base. One squadron alone, HML/A-269, claimed 90 tanks, 77 armoured personnel carriers and 156 other vehicles during the campaign. AH-1W also served in Afghanistan, however, from 2010 it began to be replaced by the AH-1Z with the final USMC aircraft being retired in late 2020.

Bell AH-1W Super Cobra
163936 was operated by VX-9 at NAS China Lake in the late 2000s. The stub wings are carrying hellfire missiles on the outer pylon and a LAU-10 pod for 5" Zuni rockets on the inboard. The chaff and flare dispenser can be seen mounted above the wing.

ATTACK HELICOPTERS

Bell AH-1Z Viper

The ultimate development of the original helicopter gunship, the AH-1Z Viper shares its engines and transmission system with the UH-1Y Huey, giving a maximum take-off weight almost two tonnes heavier than its predecessor.

While the AH-1W had served the US Marine Corps (USMC) effectively, by the late 1990s it was showing its age. As well as the poor performance of its Gen II FLIR in comparison to modern systems, the cockpit had become an ergonomic nightmare with the avionics for the radar warning receiver (RWR) and AIM-9 controls blocking the view. Meanwhile switches for the defensive aids suite, located behind the pilot, had to be accessed by feel. The answer lay in a further upgrade based in part on the unsuccessful bid for the UK's attack helicopter programme. This would become the AH-1Z Zulu Cobra or Viper.

Upgrades for the AH-1Z

Reusing some parts of the AH-1W fuselage the Viper has a new transmission system, including four-bladed rotors, capable of absorbing all the power from the T700 engines and also used on the USMC's UH-1Y Venom. An all-new AN/AAQ-30 sensor turret is mounted under the nose feeding remodelled cockpits with multifunction displays and hands-on collective and stick (HOCAS) controls.

The turrets' Gen III FLIR sensor is now able to detect targets at ranges of up to 21km (12.4 miles) compared to the more typical 5km (3 miles) of the AH-1W. The defensive aids system has been improved with four ALE-47 countermeasures dispensers and upgraded radar and optical warning systems. For carrying offensive weaponry the stub wings have been enlarged allowing four Hellfire to be carried on the inboard stations for a full load out of 16 missiles. Sidewinder missiles can now be carried on an additional wingtip station while the wings themselves hold fuel for increased range.

The Viper was declared combat ready in 2010 and saw service in Afghanistan, but by 2020 it had completely replaced the AH-1W in the USMC. Although the Marines took delivery of their last aircraft in 2022, it is being continually upgraded with new weapons and capabilities. Bahrain and the Czech Republic meanwhile are still building up their fleets, while Slovakia and Nigeria are potential future operators.

Bell AH-1Z Viper

AH-1Z serial number 169815 serves with the US Marine Corps.

AH-1Z

Dimensions: Length: 17.75m (58ft 3in); Rotor diameter: 15m (49ft 2in); Height: 4.37m (14ft 4in)
Weight: 8392kg (18,500lb) maximum take-off
Powerplant: 2 x General Electric T700-401 turboshaft engines, 1300kw (1800shp)
Maximum speed: 411km/h (255mph)
Range: 690km (430 miles)
Service ceiling: 6100m (20,000ft)
Crew: 2
Armament: 1 x 20mm (0.78in) M197 Gatling cannon, 70mm (2.75in) Hydra 70 and APKWS rockets, AGM-114 Hellfire, AGM-179 JAGM, AIM-9 Sidewinder missiles

ATTACK HELICOPTERS

Mil Mi-24

The Mil Mi-24, NATO designation Hind, design dates to the 1960s and Mil's proposal for a flying infantry fighting vehicle.

A mock-up was produced based on the unsuccessful Mil Mi-22 utility helicopter featuring a cabin for eight troops, stub wings to carry rockets or missiles and a 23mm (0.9in) cannon attached to the landing skids. Having observed the US military's use of attack helicopters in the Vietnam War, the Soviet military approved further development. This led to a revised design based on the Mi-8 Hip transport with a turret-mounted heavy machine gun under the nose. Flight testing, which commenced in 1969, saw the wings gain anhedral to address stability issues at high speed. At the same time, to increase the effectiveness of the tail rotor, it was repositioned and the direction of rotation reversed. Achieving its initial operating capability in 1971 the Mil-24A differed from later aircraft in having a slab-sided cockpit with large, flat, transparencies. This housed the pilot and gunner in tandem seats with the former's seat to the rear and offset to port, and the gunner's offset to starboard. Some aircraft also had space for a flight engineer to the rear of the cockpit. The familiar layout with separate cockpits under bubble canopies emerged with the Mi-24D, Hind D, that entered production in 1973 and has remained the layout used for all subsequent models. The flight engineer's position was moved behind the cockpits to a position ahead of the main cabin. From 1976 the Mi-24V, Hind E, entered production, proving to be the definitive version with over 1500 produced.

For battlefield use the Hind can carry up to eight troops in the cabin, who can exit through horizontally split doors on either side of the fuselage. Each stub wing has three hardpoints that can carry a variety of stores. While the outermost are limited to carrying twin launchers for AT-6 Spiral anti-tank guided missiles, the inner pylons can carry bombs, rocket pods and gun pods. The airframe meanwhile was given extensive armour protection, although events proved that this did not provide adequate defence.

First combat action

The Mi-24's first major combat operation was in Afghanistan, the Afghan government taking delivery of aircraft in April 1979 and the Soviets operating their own examples alongside them in due course. The Hind was used for helicopter assaults, in direct support of ground operations, and convoy protection. It soon became apparent that carrying troops compromised the attack capabilities, the additional weight being a handicap at the high altitudes in Afghanistan.

Consequently, the Mi-8 Hip generally operated in the troop-carrying

Mi-24D
Dimensions: Length: 21.6m (70ft 9in); Wingspan: 6.5m (21ft 4in); Height: 6.5m (21ft 4in)
Weight: 12,000kg (26,455lb) maximum take-off
Powerplant: 2 × Isotov TV3-117 turboshaft engines, 1600kW (2200shp) each
Maximum speed: 300km/h (186mph)
Range: 450km (280 miles)
Service ceiling: 4900m (16,100ft)
Crew: 2–3
Armament: 1 x 23mm (0.9in) GSh-23V twin-barrel cannon, up to 2400kg (5290lb) of ordnance including SA-18, air-to-air missiles, AT-9 anti-tank missiles and S-8 unguided rockets

Mil Mi-25D
Known as the Mi-25D, the export version of the Mi-24 is operated by several countries, including Peru. '694' belongs to Air Squadron 211 ('Dragons of the Air'), Air Group 2, Air Wing 3, Peruvian Air Force, based at Vitor in July 2012.

ATTACK HELICOPTERS

Mil Mi-24H Hind-E
This Polish operated aircraft has AT-6 anti-tank missiles on the wingtip stations while the infra-red jammer is visible on top of the fuselage. Poland is planning to replace its aircraft with AH-64E Apaches and it has sent a number of its Hinds to Ukraine for use in the conflict with Russia.

role while Hinds were used as pure attack helicopters. With the cabin clear some aircraft mounted light machine guns in the doors to allow the flight engineer to provide suppressing fire outside the arcs of the main weapons. In combat Hind formations would often operate in a 'wheel of death' pattern, attacking from multiple directions before circling the target and making additional passes as required.

While the Hind's weapon carrying ability enabled it to take the fight to the mujahideen, it was not a one-sided affair. Initially the rebels' main weapons against the Mi-24 were heavy machine guns or cannon, however, even these could present problems. The Hind's armour still left areas such as the cockpit, powerplant and transmission system vulnerable to 12.7mm (0.5in) rounds while the fuel tanks, hydraulics and the main rotor gearbox oil system also proved to be susceptible to damage. The introduction of Stinger missiles to Afghanistan in 1986 created more problems for the Mi-24 as the lack of exhaust suppression made them particularly vulnerable. In response, rather than transiting towards their targets at medium level before dropping down to attack, Hind crews would operate at low-level throughout the sortie giving SAM operators less time to react. Although an exhaust suppressor was later developed for the aircraft it is rarely seen fitted. More common is an infrared jammer mounted at the junction between the main gearbox housing and the tail boom.

A Russian-made Mi-24 helicopter from the US Army Test and Evaluation Center, Threat Support Activity, Las Vegas, Nevada, passes over to provide a simulated hostile threat during training.

ATTACK HELICOPTERS

Mi-24P

In the early 1980s the Mi-24P, Hind F, entered service. This replaced the 12.7mm (0.5in) gun with a fixed 30mm (1.18in) GSh-30-2K twin barrel cannon mounted on the starboard forward fuselage. This provided much greater hitting power at the expensive of some flexibility in aiming.

A more specialized model is the Mi-24RKhR, Hind G1, which has equipment for chemical, biological and radiation sampling, including sample gathering claws at the wingtips. These were first seen in the aftermath of the 1986 Chernobyl disaster. However, the Ukrainian armed forces appear to have reactivated at least one as a gunship post the Russian invasion of 2022.

Mi35M

Based on experience in Afghanistan and the Chechen wars a number of upgrades were proposed for the Hind, which can be retrofitted to earlier aircraft. These include replacement of the rotor system with the main rotor head and composite blades from the Mi-28 together with its noise reducing X-shaped tail rotor. Airframe alterations include a new shorter wing that stops at the second pylon and a fixed undercarriage that improves crash survivability for a modest reduction in top speed.

Weapons and self-defence upgrades are also available including a more powerful 23mm (0.9in) cannon to replace the 12.7mm (0.5in) gun and the latest laser-guided AT-9 Spiral 2 anti-tank missiles. In the air-to-air role the SA-18 Grouse can be fitted for counter UAS and helicopter use.

Finally, a modern avionics suite can be fitted providing an NVG compatible cockpit, a digital weapons control system and a modern electro-optical system, which gives an all-weather day/night capability. New build aircraft incorporating all these features are designated Mi-35M and use a more powerful Klimov VK-2500 turboshaft engine, while aircraft that have received the systems upgrades but retain the original airframe, rotor system and engines are known as Mi-35P. The Mi-35P has a top speed of 335km/h (208mph) compared to the 300km/h (186mph) of the Mi-35M, however, the M models have a 300m (984ft) higher service ceiling and greater payload thanks to their lighter structure.

Russia–Ukraine War

The Hind has been active in the Russian invasion of Ukraine, where both sides operate variants of the type. The widespread presence of surface-to-air missiles, however, has taken a toll, predominantly on the Russian forces, who have reportedly lost 18 Hinds versus seven of Ukraine's. This has forced a change in tactics, with Hinds flying towards their target at low level before pulling up to launch a salvo of rockets in the climb before turning away and dropping back down to tree-top height. This does little for the unguided weapons' accuracy, but does seem to have stemmed the aircraft's loss rate.

Robust, simple to maintain and with more than 2500 examples built, the Hind remains in service with over 50 countries and is likely to be in operation for decades to come.

Mil Mi-25D
'854' from 1335 Squadron, Libyan Air Force, based at Mitiga, in October 2009. This aircraft was reportedly lost and its pilot killed on 9th April 2011, while flying for the Free Libyan Air Force during the Libyan Civil War.

ATTACK HELICOPTERS

Mil Mi-35P

35206, belonging to 714th Anti-armour helicopter squadron ("Shadows") of the 98th Aviation Brigade, Serbia Air Force and Air Defence, based at Ladevci Air Base, September 2024. This aircraft was part of an order of 11, purchased from the Cyprus National Guard and overhauled by UTVA at Pancevo.

Mil Mi-35M

The Mil Mi-35M is to date the most advanced version of the 'Hind' series to have entered serial production, featuring updated avionics, targeting, and communication systems. As well as serving with the Russian Aerospace Forces, the Mi-35M has been exported; '03 Red' serves with the Air Defence Forces of Kazakhstan.

ATTACK HELICOPTERS

Mil Mi-28N

Although the Mi-28 lost to Kamov's Ka-50 in the competition to provide Soviet forces with an attack helicopter, in 1991 the Russian Ministry of Defence tasked Mil with developing a night capable aircraft, the Mi-28N, known to NATO as Havoc-B.

Building on Mil's experience with the Mi-24 Havoc-B features a stepped cockpit arrangement for the pilot and weapons system operator; unlike the Hind there is no cabin as such although there is a compartment capable of holding two passengers. Intended for rescuing aircrew from the battlefield this is primarily an avionics compartment and is accessed via a door just below the left-hand engine's exhaust. A conventional rotor layout is used with a five-bladed main rotor and four-bladed tail rotor in an X configuration similar to that on the AH-64. The engines are mounted to either side of the fuselage above stub wings, which each have two pylons for a range of stores. A 30mm (1.18in) cannon is carried underneath the forward fuselage on a flexible mount able to aim to 110° either side of the centreline. For the night attack role, the Mi-28N gained an under nose combined TV and forward looking infrared (FLIR) sensor that can traverse through the same range as the cannon; above this is a ball turret containing a laser designator. Survivability is enhanced with a fully armoured cockpit surrounded by 10mm (0.39in) aluminium alloy armour able to withstand 12.7mm (0.5in) rounds and undercarriage and seats designed to absorb a 12m/s vertical impact. Additional protection is provided by a 42mm (1.65in) thick armoured glass windscreen, also able to survive hits from 12.7mm (0.5in) rounds, and 22mm (0.8in) thick side windows rated for 7.62mm (0.3in) rounds.

The underwing pylons have four hardpoints able to carry a range of weapons including four-round AT-9

Mi-28N

'04 White' features the overall grey finish applied to some Mi-28s, a finish that makes it look even more like the Boeing AH-64 Apache it superficially resembles.

ATTACK HELICOPTERS

Spiral-2 launchers, the anti-tank guided missile having a range of 5.8km (3.6 miles) and a warhead that can defeat up to 850mm (33in) of explosive armour. Alternatively, unguided weapons include 80mm (3.3in) S-8 rockets in 20 round launchers, S-13 122mm (4.8in) rockets in five round launchers and KMGU-2 mine dispenser pods. For an anti-air capability launcher, packs for SA-18 Grouse air-to-air missiles can be fitted. The defensive aids suite features radar and laser warning receivers to alert the crew if they are being targeted and an ultra-violet warning receiver to detect a missile launch. Chaff and flares are dispensed from pods mounted on the wingtips while an infrared jammer is carried beneath the fuselage.

Grounded after first flight

First flight of the Mi-28N took place in November 1996, however, shortly after an issue was found with the new main rotor gearbox and the aircraft was grounded. Due to the limited funding for the programme it did not return to the air until 2002. Although the first Mi-28N featured a mast-mounted radar the second Havoc-B that followed in 2004, and all subsequent examples, did not have it as development of the system was ongoing. They do, however, have new main rotors designed to absorb hits from 30mm (1.18in) rounds, and improved engine and fuel controls. The Mil bureau's faith in their design finally paid off in 2005 when the Russian Ministry of Defence placed an initial order for 67 aircraft with order for a further 100 being placed by 2010.

The Mi-28N first saw combat during the Syrian Civil War in 2016 with Russian forces supporting the Syrian Army in recapturing the city of Tadmur from Islamic State. Havocs employed unguided rockets and AT-9 anti-tank guided missiles to destroy enemy positions.

Mi-28 have been used extensively in the Russian invasion of Ukraine although this has not been without loss. Pre-war doctrine was based on

Mil Mi-28N

Dimensions: Length: 17.91m (58ft 9in); Rotor diameter: 17.2m (56ft 5in); Height: 3.82m (12ft 6in)
Weight: 11,500kg (25,353lb) maximum take-off
Powerplant: 2 × Klimov VK-2500 turboshaft engines, 1636kW (2194shp) each
Maximum speed: 320km/h (200 mph)
Range: 435km (270 miles)
Service ceiling: 5700m (18,700ft)
Crew: 2
Armament: 1 × 30 mm (1.18in) Shipunov 2A42 cannon on steerable mounting, AT-9 anti-tank guided missiles, SA-18 air-to-air guided missiles, S-8 and S-13 unguided rockets, KMGU-2 mine dispensers and gun pods

Mi-28N
On display at Kubinka in 2015, '52 Yellow' is finished in the standard tactical camouflage scheme worn by the Russian Mi-28 fleet.

ATTACK HELICOPTERS

attack helicopters approaching the target area at low-level and high speed to minimize the risk of detection. The aircraft then pop-up to acquire their target, delivering their weapons in a shallow dive. Unfortunately, this has not survived the trial of combat, and 11 Havocs were lost in the first year of the conflict likely due to the proliferation of man-portable surface-to-air missiles. In response Russian units have taken to remaining on their side of the frontline while targeting troops and vehicles on the Ukrainian side.

Further developments

Development of the Mi-28 continued with the Mi-28NM ordered in 2017. This has an updated avionics suite, electro-optical turret and defensive aids system, and adds a new N025E radar above the main rotor. This can detect bridges at 20km (12.4 miles) and main battle tanks at 10km (6.2 miles) and has led to the removal of the thimble-shaped nose radome.

While Russian Aerospace Forces are the main operators of the Havoc, Algeria placed an order for 8 of the Mi-28NE export variant in 2012, which was subsequently increased to 42 in 2016. Iraq also operates the Mi-28NE with 15 having been delivered from 2013.

A Mi-28N at Kubinka Air Force Base in 2015. As well as the nose mounted sensors this view also clearly shows the design of the tail rotor with two pairs of rotor blades stacked together.

ATTACK HELICOPTERS

Changhe Z-10

After the 1988 transfer of the People's Liberation Army Air Force's helicopters to the People's Liberation Army Ground Force (PLAGF) and the establishment of its aviation arm, it was decided to acquire a dedicated attack helicopter.

Development began in 1995 with the Russian Kamov Bureau commissioned to develop a preliminary design in the six-tonne weight class, the same as the Eurocopter Tiger. This design was then passed to the Aviation Industry Corporation of China (AVIC) for further development with Changhe Aircraft Industries Corporation (CAIC) assigned to serial production.

Fiery Thunderbolt

Although the Z-10, also known as Fiery Thunderbolt, follows the standard attack helicopter layout the fuselage and wings are blended giving a less utilitarian appearance than its contemporaries, while the fuselage itself has a hexagonal cross section. The stub wings each mount two hardpoints for external weaponry. The fuselage sides slope inwards from a distinctive chine that runs along the mid-fuselage, increasing the internal volume. Power is provided by two 1200kW (1600shp) Zhuzhou WZ-9 turboshaft engines, replacing the 1142kW (1530shp) Pratt & Whitney PT6C used in the prototype due to US export restrictions, which Pratt & Whitney were found guilty of violating. The engine exhausts initially vented to the sides, but are now directed upwards into the rotor downwash to mix the hot gases with ambient air reducing the infrared signature.

As well as an electro-optical sensor system in the nose turret the Z-10 can be fitted with a mast-mounted millimetre band radar. A comprehensive defensive aids suite is fitted while aircraft have been seen with additional armour plate mounted over the cockpit and engine sides. As well as the 20mm (0.78in) cannon mounted under the nose the Z-10 can carry laser and radar-guided missiles. Static display aircraft have also been seen with loitering munitions and torpedoes, the latter being unusual for an attack helicopter but would allow them to assist ASW operations if deployed with a task group. It is also reported the Z-10 has a load lifting capability that again would be unusual but potentially of use when supporting special forces.

Now in widespread service with the PLAGF, the Z-10 is stationed in the Taiwan Strait and the Tibetan plateau where its high-altitude performance is crucial. To date the only other operator of the Z-10 is the Pakistan Army Aviation Corps, which expects to have taken delivery of 40 by 2026.

Z-10K

Dimensions: Length: 14.2m (46ft 7in); Rotor diameter: 13m (42ft 8in); Height: 3.85m (12ft 8in)
Weight: 7200kg (15,873lb) maximum take-off
Powerplant: 2 x WZ-9C turboshaft engines, 1200kW (1600shp)
Maximum speed: 290km/h (180mph)
Range: 800km (500 miles)
Service ceiling: 6400m (21,000ft)
Crew: 2
Armament: 1 x 23mm (0.9in) PX-10A cannon, 57mm (2.2in), 70mm (2.75in) and 90mm (3.5in) rockets, HJ-8, HJ-9, AKD-9, and AKD-10 laser-guided missiles, AKD-21 radar-guided missiles, and GB25 and GB50 guided bombs

---

Changhe Z-10K

This Z-10K belongs to the 4th (Rotary Wing) Brigade of the PLAAF, based at Huangpi. It wears a three colour tactical camouflage and carries AKD-10 anti-tank missiles under its stub wings.

ATTACK HELICOPTERS

HAL Prachand

The HAL Prachand's development was spurred by the 1999 Kargil War when the Indian Armed Forces lacked an attack helicopter capable of operating at high Himalayan altitudes. In 2006 HAL started development with a design based on its own Dhruv light utility helicopter.

The resulting aircraft has a narrow fuselage with tandem seating for the crew of two. The hingeless main rotor and bearingless tail rotor are shared with the Dhruv, as are the HAL/Turbomeca Shakti-1H1 engines. These combine to give the Prachand excellent manoeuvrability and high-altitude performance with a service ceiling in excess of 6500m (21,000ft). Unlike the Dhruv, the Prachand has a crashworthy wheeled under carriage, self-sealing fuel tanks and armour protection.

Armament is provided by a nose mounted 20mm (0.78in) M621 cannon and four wing-mounted hardpoints able to carry a range of air-to-air and air-to-surface missiles. These are operated via the helicopter's target acquisition and designation system and its integrated helmet-mounted sights. A datalink is fitted to allow target information to be exchanged with other platforms while the nose turret includes an infrared sensor and a laser designator.

Operational testing began in 2020 with the Indian Air Force (IAF) deploying two aircraft to the mountainous Ladakh region near the border with Pakistan. The first of seven planned Indian Army Aviation Corps squadrons was formed in June of 2022 with the first IAF squadron stood up at Jodhpur that October. These aircraft will be operated alongside India's recently purchased AH-64E in a light-heavy mix. HAL meanwhile is in the process of selling the Prachand to Nigeria, with potential sales to Argentina, Egypt and the Philippines.

HAL Prachand

ZF 4834, Indian Air Force, assigned to No.143 Helicopter Unit (Dhanush: 'Bow') based at Jodhpur Air Force Station, India, in October 2022. It is one of the initial operational aircraft and carries the word Prachand, which is usually translated as "fierce", on the nose.

HAL Prachand

Dimensions: Length: 15.8m (51ft 10in); Rotor diameter: 13.2m (43ft 4in); Height: 4.70m (15ft 5in)
Weight: 5800kg (12,787lb) maximum take-off
Powerplant: 2 × HAL/Turbomeca Shakti-1H1 turboshaft, 1032 kW (1384shp) each
Maximum speed: 280km/h (174mph)
Range: 700km (435 miles)
Service ceiling: 6500m (21,300ft)
Crew: 2
Armament: 1 x 20mm (0.78in) M621 cannon, FZ275 laser-guided rockets, Dhruvastra anti-tank and Mistral air-to-air missiles, guided and unguided bombs

ATTACK HELICOPTERS

Harbin Z-19

Also known as Xuanfeng, or Black Whirlwind, the Harbin Z-19 is a light reconnaissance and attack helicopter based on the Harbin Z-9 utility helicopter, which is itself a licence-built copy of the Eurocopter AS365 Dauphin.

Although the engines, rotors and transmission have been carried across from the Z-9WA, the fuselage has been substantially modified. The crew are, typically for an attack helicopter, sat in tandem giving a much slimmer frontal profile while the cabin has been completely removed. At the same time armour plating and crash-resistant seats were added to improve survivability.

The most obvious similarity with the Z-9 is the Fenestron tail rotor, enclosing the blades in a duct in the large vertical stabilizer. This configuration is less susceptible to damage when operating at low level and in confined areas, while also having a lower noise footprint due to the reduction in the strength of the tip vortices. Despite these advantages it has not proved a popular configuration for attack helicopters due to the greater weight and hover power requirements.

Harbin Z-19
LH952506 belonging to the 82nd Army Aviation Brigade of the Central Theatre Command from China's People's Liberation Army, based at Baoding Airfield, Hebei, China, in December 2023.

With an emphasis on the reconnaissance role, the Z-19 does not feature an integral cannon although there are four underwing hardpoints for guns, rockets and missiles. There is a nose-mounted sensor pod with visual and infrared cameras as well as a laser rangefinder while some aircraft have been seen with a mast-mounted millimetre band radar, again emphasizing the reconnaissance role.

Entering service with the People's Liberation Army Ground Force in 2012 over 180 Z-19 have been delivered. Although an export version has been promoted by Harbin, to date there have been no foreign sales.

Harbin Z-19
Dimensions: Length: 12m (39ft 4in); Rotor diameter: 11.93m (39ft 2in); Height: 4.01m (13ft 2in)
Weight: 4250kg (9370lb) maximum take-off
Powerplant: 2 x WZ-8C turboshaft, 700kW (940shp) each
Maximum speed: 280km/h (174mph)
Range: 700km (435 miles)
Service ceiling: 6000m (19,685ft)
Crew: 2
Armament: Rocket, gun and cannon pods, HJ-8 air-to-surface and TY-90 air-to-air missiles

RECONNAISSANCE AND UTILITY

Proving the flexibility of helicopters, these aircraft can be adapted to a variety of roles. Although often broadly designed around one capability, such as the medium transport UH-60 Black Hawk, depending on the operator's requirements they can be adapted with sensors and weapons for reconnaissance and light attack. The addition of a hoist also gives most of the types listed here a search and rescue ability.

This chapter includes the following helicopters:

- UH-60 Black Hawk
- Bell UH-1
- Airbus H145M
- NHIndustries NH-90
- Bolkow Bo105
- Bell OH-58 Kiowa
- Bell CH-146 Griffon
- Eurocopter AS365 Dauphin II
- Aérospatiale SA 342 Gazelle
- Bell 412
- Changhe Z-11WB
- Kamov Ka-226
- Bell UH-1Y Venom
- Harbin Z-9
- Harbin Z-20
- Eurocopter AS532 Cougar

Left: A UH-60 Black Hawk assigned to 82nd Combat Aviation Brigade, 82nd Airborne Division, lifts off after refueling, 2018.

RECONNAISSANCE AND UTILITY

UH-60 Black Hawk

Drawing on its experiences in Vietnam, in 1972 the US Army released its requirements for a replacement for the UH-1 Huey, which was to become the UH-60.

The Army required an aircraft powered by two 1210kW (1645shp) General Electric T700 turboshaft engines that were under development as part of the Utility Tactical Transport Aircraft System (UTTAS) programme. Other requirements included a payload of 11 troops or 1200kg (2645lb), a cruise speed of 270km/h (167mph) and an endurance of at least 2.3 hours. The payload was chosen to preserve squad integrity and minimize the number of aircraft in the landing zone, while the speed and endurance levels were aimed at maintaining mission integrity and minimizing time lost to refuelling. A further limitation was a requirement to transport the aircraft in a C-130 Hercules with minimal disassembly, which would constrain the height of the resulting aircraft and drive some novel design features. Boeing-Vertol and Sikorsky both built prototypes for a fly-off that took place during 1976. Sikorsky's UH-60 was selected as the winner in December of that year with deliveries of production aircraft starting in late 1978 for service entry with the 101st Airborne Division the following June.

Unusual design solution

To fit in the cargo hold of a C-130, the UH-60 originally had the main rotorhead just a few centimetres above the gearbox housing, however, this led to unacceptable vibration levels. This was remedied by raising the rotor 38cm (15in) using a shaft extender that can be removed for air transportation. Meeting the overall length limitations also meant an unusual design solution with the main rotor mounted ahead of the centre of gravity to ensure rotors of sufficient length could still be used. To counter this the tail rotor is canted over by 20° to provide 180kg (400lb) of lift aft of the centre of gravity. At the same time the large horizontal stabilizer provides additional trim moving in response to the aircraft speed to maintain a broadly level attitude. Tragically at least five Black Hawks were lost during the 1980s when high intensity radiated fields (HIRF) interfered with the stablizer's fly-by-wire system, causing uncommanded stabilator movement. HIRF are associated with radio broadcasting antenna and the US Army's initial action was to restrict

Sikorsky HH-60G Black Hawk
Pave Hawk 26471 was operated by the 210 Air Rescue Squadron of the Alaska Air National Guard during which it was regularly seen with skis fitted to the undercarriage for snow landings. It has since transferred to European theatre.

HH-60G
Dimensions: Length: 19.76m (64ft 10in); Rotor diameter: 16.36m (53ft 8in); Height: 5.13m (16ft 10in)
Weight: 9980kg (22,000lb) maximum take-off
Powerplant: 2 × General Electric T700-GE-701C/D turboshaft engines, 1418kW (1902shp) each
Maximum speed: 294km/h (183mph)
Range: 933km (580 miles)
Service ceiling: 4270m (14,000ft)
Crew: 4
Armament: 2 x 7.62mm (0.3in) M240 machine guns or 2 x 12.7mm (0.50in) GAU-19 gatling guns

RECONNAISSANCE AND UTILITY

S-70A-9 Black Hawk
Originally delivered to the Royal Australian Air Force, A25-101 was built by Sikorsky while the rest of the fleet were produced by de Havilland Australia. Retired in 2021 they are being replaced by UH-60M models.

operations near the several hundred HIRF sites identified worldwide. The US Navy's Seahawk aircraft did not suffer this issue having been designed to operate near ship's radars and radios with shielding around the fly-by-wire system's sensitive components. Army Black Hawks eventually had similar shielding fitted that enabled the operating restrictions to be relaxed.

Lessons from Vietnam
Survivability lessons from the Vietnam War were incorporated into the UH-60's design. These included armoured crashworthy seats and an energy absorbing undercarriage to minimize the impact of a heavy landing on the occupants. Controls and drive shafts are tolerant to ballistic impacts and where necessary duplicated. The main rotor blades, featuring titanium spars, and fuselage are designed to absorb hits from 23mm (0.9in) rounds while the large vertical stabilizer provides directional stability in the event of tail rotor loss. In the event of a crash the fuel tanks are self-sealing and crashworthy while the fuselage can retain the gearbox and engines in the event of a crash protecting the cabin occupants. To reduce the likelihood of being hit in the first place the Black Hawk is fitted with a Hover Infrared Suppressor System (HIRSS), which mixes cool air with the engine exhaust decreasing the range at which it can be detected by heat-seeking missiles.

The External Stores Support System (ESSS) mounted above the cabin doors allows the carriage of up to 2270kg (5000lb) of stores on either side on two hardpoints. Made of composites they can be fitted as required for the mission but require the necessary attachment points and electrical connections to be installed in the airframe. So equipped a Black Hawk can carry up to 16 Hellfire missiles with reloads in the cabin; alternatively Maverick and Sidewinder missiles can be carried or 70mm (2.75in) rockets in multi-round pods. Alternatively, four fuel tanks can be fitted giving an unrefuelled range of over 2000km (1100nm). Cabin-

RECONNAISSANCE AND UTILITY

mounted 7.62mm (0.3in) and 12.7mm (0.5in) guns can be fitted at the crew stations aft of the cockpit while the M230 30mm cannon can be carried on the ESSS.

Operational use

The first operational use of the UH-60A was during the 1983 invasion of Grenada where they were used to deliver SEAL Team 6, Delta Force and Army Rangers to their objectives. This was followed in 1989 by the invasion of Panama. Although the UH-60 was heavily used throughout the 1991 Gulf War, one action is particularly noteworthy. On 24 February, the first day of the ground campaign, the 101st Airborne Division used over 60 Black Hawks and 30 Chinooks to move to a forward operating base (FOB) 130km (80 miles) inside Iraq. This was the largest airborne assault in history with Apache and Cobra attack helicopters adding to the numbers. It was followed by a second lift the following day taking the 101st another 250km (155 miles) into Iraq and within 230km (142 miles) of Baghdad. The Gulf War also saw use of the Black Hawk in the medevac role with a rack for six stretchers replacing the troop seats.

From the late 1980s the improved UH-60L entered service, which used the more powerful 1385kW T700-GE-701C engines and uprated gearbox that had been developed for the SH-60B Seahawk. This was vital to counter the increase in weight that had come during the 1000 aircraft production run of the UH-60A. This was down to the addition of systems such as HIRSS, ESSS fittings and the rotor de-icing system, which meant a late production aircraft was around 500kg (1100lb) heavier than the first ones off the line. As well as restoring performance, the extra power improved the Black Hawk's capabilities with the load lifting capacity increased from 3628kg (7204lb) to 4082kg (8999lb) therefore increasing the range of artillery and vehicles that could be carried.

This was followed in the early 2000s by the UH-60M with 1500kW (2040shp) T700-GE-701D engines that are coupled to new composite main rotor blades with an increased chord and swept anhedral tips to provide improved performance at high weights and altitudes. Together these increased the maximum all up mass by 450kg (992lb) allowing a fourth crew member to be permanently carried and the door guns to be manned on both sides at all times. The fuselage structure has been improved with a reduction in part count to reduce cracking and corrosion hotspots, and the transmission support beams have also been strengthened. A cockpit overhaul introduced four colour multifunction displays, based on configurations developed for foreign sales. As well as reducing the pilots' workload by simplifying the display of flight critical information, the upgrade

Sikorsky HH-60G Pave Hawk
Designed for the Combat Search and Rescue role, the Pave Hawk has a number of modifications, including air-to-air refuelling. This example operated by the US Air Force Reserve Command saw service in Afghanistan.

RECONNAISSANCE AND UTILITY

has also allowed for the addition of systems such as ADS-B and RNAV, which are almost mandatory for operations in civilian airspace. Other improvements include a new HIRSS that directs the exhaust gases into the rotor downwash, along with laser and radar warning receivers, infrared jammers and decoy dispensers. UH-60M deliveries started in 2006 to replace the oldest US Army Black Hawks while an upgrade package for the UH-60L replaced the original cockpit with a similar arrangement to that in the UH-60M, which are designated UH-60V. UH-60M first saw operational service in Iraq in 2008, followed by the HH-60M air ambulance in 2010.

As an indication of the improvement in performance the UH-60M has a maximum all-up mass some 750kg (1653lb) more than that for the UH-60A at 9980kg (22,000lb). The new aircraft also has a higher cruise speed at 283km/h (175mph) to the originals 260km/h (161mph) and at a weight of 8165kg (18,000lb) the UH-60M can climb vertically at 5m/s (994in/

Black Hawk helicopters delivering U.S. Army personnel to the passenger terminal at Sather Air Base, Iraq, in June 2008.

min) while the A model would have to maintain forward speed to climb.

Black Hawk variants

Specialized versions of the Black Hawk date back to 1980 when the EH-60A Electronic Warfare (EW) variant was developed. Also known as Quick Fix, this featured an array of dipole antenna on the sides of the tail boom while a retractable whip aerial folded down from the underside of the fuselage. Designed for communications jamming, 66 EH-60A/C were built.

RECONNAISSANCE AND UTILITY

Soldiers from Company C, 1st Battalion, 207th Aviation Regiment conduct air assault training out of Wheeler Army Air Field in Wahiawa, Hawaii, June 2015. The drills were conducted on the island of Oahu using Sikorsky UH-60 Black Hawk helicopters.

The UH-60Q DUSTOFF is a medical evacuation derivative of the UH-60L with a comprehensive onboard medical suite. To enable rescues to be carried out in poor weather, a radar and FLIR are fitted to the nose. The medevac role has subsequently been filled by the HH-60L and HH-60M. Multiple special operations versions have been developed, the latest being the MH-60M used by the 160th Special Operations Aviation Regiment (SOAR). This is equipped with an in-flight refuelling probe, AN/APQ-187 terrain-following radar and upgraded defensive systems. These can be used either as assault aircraft carrying troops or as a Direct Access Penetrator (DAP) with a full load of weapons on the ESSS to provide fire support. The US Air Force also operate a Black Hawk variant, the HH-60G Jolly Green II and HH-60W Pave Hawk, both of which are used for combat search and rescue and special forces insertion and extraction. The US Navy meanwhile, as well as operating the SH-60 (see separate entry), also acquired the MH-60S Knight Hawk to replace its CH-46 Sea Knights. This pairs the fuselage of the UH-60L with the folding rotors, engines and transmission system from the SH-60 family. With over 5000 Black Hawks produced to date, the Hawk family is operational with over 40 countries and will remain in service for decades to come.

Sikorsky UH-60A Black Hawk

Based at Chievres in Belgium, this A model Black Hawk was used by the Supreme Headquarters Allied Powers Europe (SHAPE) whose crest is seen on the cargo door.

Bell UH-1

One of the most successful helicopter designs of all time, the Bell UH-1 began life in response to a late 1950s US Army requirement for a battlefield utility and medical evacuation aircraft.

Bell's proposal was powered by a 520kW (700shp) Lycoming T-53 gas turbine placed above and behind the cabin. The cabin itself was easily accessed via large sliding doors on either side above the landing skids. Designated the HU-1A and officially named Iroquois, an initial production contract for 182 aircraft was awarded in 1959. These aircraft used the more powerful 570kW (770shp) version of the T53, addressing concerns from service tests that it was underpowered. The first units to field the HU-1A were the 101st Airborne Division, 82nd Airborne Division and the 57th Medical detachment with this unit arriving in Vietnam in March 1962, beginning the type's long association with the country.

Enduring nickname

Nicknamed 'Huey' due to the original HU-1 designation, the US Army was requesting improvements even as the A models were being delivered. This resulted in the HU-1B with the T53 producing 720kW (960shp) and a longer cabin providing seating for seven passengers or four stretchers. While deliveries of the HU-1B began in 1961 work had already begun on the C model, primarily intended to improve the aircraft's performance while armed with external rocket and gun pods. As well as increasing the T53's output to 820kW (1100shp) the rotor system was improved with a larger diameter main rotor. This also required a longer tail boom and an increase in the area of the horizontal stabilizer, which was also mechanically linked to the cyclic, altering its pitch in flight to allow a wider acceptable range for the centre of gravity. Early experience in Vietnam also led to improvements to the engine intake filter system to deal with dusty conditions.

In a search for further capacity Bell would again stretch the fuselage resulting in what would be the definitive Huey configuration. First flying in August 1961, the HU-1D was 1.04m (41ft) longer than the B model allowing seating for up to 14 troops or six stretchers. The sliding doors were enlarged while hinged doors were fitted at the forward end

Bell UH-1B

An early foreign operator of the Huey, the Royal Australian Air Force acquired A2-1021 in 1964 through the Foreign Military Sales (FMS) route. Serving with 9 Squadron in Vietnam it would be sold into civilian use in 1990.

Bell UH-1H

Dimensions: Length: 17.62m (57ft 10in); Rotor diameter: 14.6m (48ft 3in); Height: 4.41m (14ft 6in)
Weight: 4309kg (9500lb) maximum take-off
Powerplant: 1 × Lycoming T53-L-13 turboshaft, 1000kW (1400shp)
Maximum speed: 204km/h (127 mph)
Range: 511km (318 miles)
Service ceiling: 3800m (12,600ft)
Crew: 1–4
Armament: Various including 7.62mm (0.3in) machine guns, 70mm (2.75in) Hydra 70 unguided rockets, TOW missiles

RECONNAISSANCE AND UTILITY

of the opening, all of which could be removed if needed. The main rotor span was increased from the 13m (42.6ft) diameter of the HU-1C to 14.6m (50ft) requiring a further extension of the tail boom. Initially powered by the 820kW (1100shp) version of the T53, by 1966 this had been uprated to 1000kW (1400shp) in the UH-1H, almost double that of the HU-1A. Although the UH designation was adopted in late 1962 and applied to all Huey models, the nickname endured.

Distinctive sound

What all these models, and subsequent ones, had in common was the distinctive thumping noise they made in flight and immortalized in many films about the Vietnam War. This was caused by the two-bladed rotor design, which was an evolution of that used on the earlier Bell 47. Mechanically simpler than a multi-bladed design, the UH-1's wide chord blades have high inertia allowing an improved safety margin in the event of engine failure. This mechanical simplicity, however, comes at the cost of increased vibration due to the greater variation in lift as the blades complete a revolution, being at a minimum in forward flight when the blades are positioned fore and aft. The thumping noise is also exacerbated in descending flight making it difficult for a Huey to make a covert approach.

The UH-1D was deployed to Vietnam from 1963 where the type was used in various roles. Aside from air assault and general transportation duties, the Huey was soon being operated in a gunship role, initially via local modification in theatre with rocket and gun pods. The UH-1C introduced a purpose-built gunship and they regularly operated in hunter-killer teams with OH-6A or OH-58A observation helicopters, these smaller aircraft operating at low-level to identify targets before the Huey made an attack run. Although the AH-1 Cobra later took over in the gunship role the UH-1 was used to successfully trial the TOW-guided missile system during the closing stages of the war.

The US Navy (USN) operated UH-1B gunships in light attack helicopter squadrons to wowrk with river patrols. These would operate in pairs from LSTs (landing ship tank) alongside 10 patrol boats and a SEAL detachment. The US Air Force (USAF) also acquired the UH-1F to support special operations in Vietnam, modifying it to use the T-58 gas turbine for commonality with the Air Force's HH-3 rescue helicopters. By 1970, however, the USAF operated the UH-1N Twin Huey in Vietnam.

Powerplant

Initially developed to fulfil a Canadian requirement, the UH-1N was powered by a Pratt & Whitney Canada PT6T turboshaft engine. This 'twin-pac' system actually consisted of two PT6A engines mounted side-by-side and feeding into a combining gearbox, the output from this then fed into the Huey's original main rotor gearbox. Although capable of providing a maximum of 1342KW (1800shp) of power the main purpose of the installation is to improve safety rather than increase all out performance.

In the event of one engine failing the other can provide up to 671kW (900shp) enabling a safe recovery to be made. After the USAF took delivery

RECONNAISSANCE AND UTILITY

Agusta-Bell AB-204
Produced under licence by Agusta in Italy, the AB-204 was used by the Italian Army and Air Force before being replaced by the longer AB-205 model equivalent to the HU-1D.

the USMC and USN received the UH-1N in 1971, again deploying them to Vietnam. The UH-1N remained in service with the USMC until 2014 having undertaken its last combat deployment in 2010 to Afghanistan.

USAF UH-1N was used to guard nuclear missile silos in the Continental USA well into the 21st century with their retirement only beginning in 2022.

While the US Army did not acquire the Twin Huey it continued to operate the single-engine variants until 2016. They saw service in Grenada, Panama and the Gulf War where over 400 aircraft were deployed alongside the UH-60 Black Hawk that was gradually replacing it.

Export service
Although the first-generation Huey is in the twilight of its career with the United States, it is still in active military service with around 40 other countries from South America to the Far East. Surplus UH-1s are also in widespread use with government agencies and commercial organizations. Somewhat ironically, until the mid-2010s the Vietnamese People's Air Force was a UH-1 operator with a collection of aircraft that had been left behind after the war.

As the UH-60 family have taken over from the Huey in US service the venerable UH-1 would still play a crucial role in SAR and other secondary roles. This US Marine Corps example is conducting a training mission in Arizona in 2006.

45

RECONNAISSANCE AND UTILITY

Airbus H145M

The H145 was the result of a joint development between Eurocopter and Kawasaki, building on the success of their earlier work on the BK117.

The new aircraft essentially combined the previous aircraft's rear fuselage with a cockpit based on the EC135 including its glass cockpit. Despite the outward similarity the H145 has a greater range and payload than its predecessor, more usable cabin area, and lower vibration and noise levels. The H145M designation covers military derivatives of the basic aircraft, which can encompass a broad range of modifications from the simple addition of UHF radios to integration of weapons and advanced sensors, depending on the operators' requirements.

The largest user of the H145M is the US Army, which operates over 400 as the UH-72 Lakota, which are manufactured in Airbus' facility in Columbus, Mississippi. Used for transportation and medevac duties in Continental USA, the UH-72 allows the more expensive Sikorsky UH-60 to be freed up for less permissive environments. Although also used as a training aircraft since 2021, it has proved less than satisfactory in the role

Airbus H145M

Operated by Escuadrón de Combate 2211 "Cobras" of the Ecuadorian Air Force, FN1253 is based at Base Aerea Simon Bolivar in Guayas province. This was the third of six H145Ms delivered to the Ecuadorian Air Force between October 2020 and April 2021.

and the Army is already looking for a replacement.

Unusually while all H145M are twin-engined, the earlier aircraft feature a four-bladed hingeless main rotor with a conventional tail rotor while later models have a five-bladed main rotor and a Fenestron. In US Service these are designated the UH-72A and UH-72B respectively.

Aside from the US Army, the German armed forces are the largest operator of the type using them for special forces support, and search and rescue, and are in the process of acquiring 62 in a light attack helicopter variant. Thailand, unusually, operates European-manufactured H145M alongside US-made UH-72A aircraft.

H145M
Dimensions: Length: 13.03m (42ft 9in); Rotor diameter: 11.0m (36ft 1in); Height: 3.45m (11ft 4in)
Weight: 3585kg (7904lb) maximum take-off
Powerplant: 2 × Turbomeca Arriel 1E2 turboshaft engine, 550kW (738shp) each
Maximum speed: 268km/h (167 mph)
Range: 685km (426 miles)
Service ceiling: 4000m (13,100ft)
Crew: 1–2
Armament: Various including 7.62mm (0.3in) machine guns, 70mm (2.75in) Hydra 70 unguided rockets, TOW missiles

NHIndustries NH90

NHIndustries was formed by companies from France, Germany, Italy and the Netherlands specifically to design and develop the NH90 utility and maritime helicopter on which work began in 1992.

This required an aircraft able to operate from sea level to 6000m (20,000ft) in temperatures ranging from -40°C to 50C°, including flight in icing and instrument flight conditions. For maritime operations the aircraft had to be able to operate from frigates and destroyers in up to a sea state six. Both a tactical transport helicopter, the TTH, and a maritime version, the NATO Frigate Helicopter (NFH), were to be produced based on the same basic airframe. Development was rapid, with assembly of the first prototype beginning in 1994 for roll-out the following year and a first flight in December 1995. Eight years later, the third prototype became the first helicopter to fly solely with fly-by-wire controls and no mechanical back-up.

Aside from the lack of mechanical controls, the NH90 features an integrated avionics suite with a glass cockpit based

NHI NH90 TTH
Operated by the Italian Army, EI-203 was among the first NH90 Tactical Transport Helicopters delivered and was seen at the Farnborough Air Show in 2008.

around five multifunction displays. A four-axis autopilot eases the pilot workload, allowing the NFH variant to operate with a crew of only one pilot, a tactical co-ordinator and a sensor operator. The fuselage is constructed largely from composites, which reduces the number of parts and the weight and in turn leads to a 30 per cent increase in endurance compared to a conventional metallic structure. While small enough to fit in a frigate hangar, or to be air transported by an A400M, the NH90 TTH can accommodate 20 troops or 12 stretchers with loading via sliding side doors or a rear ramp. The NH90 is claimed to have the lowest radar signature in its class due to the external shape and extensive use of composites. The infrared signature is minimized by suppressors on the exhausts of the two RTM322 engines while self-protection is provided by a missile launch detection system and chaff and flare dispensers. If these measures are unsuccessful in preventing the aircraft from being hit, all the important systems feature redundancy while the crew and passengers are protected by armour plating.

NH90
Dimensions: Length: 16.13m (52ft 11in); Rotor diameter: 16.3m (53ft 6in); Height: 5.31m (17ft 5in)
Weight: 10,600kg (23,369lb) maximum take-off
Powerplant: 2 × Rolls-Royce Turbomeca RTM322-1/9 turboshaft engines, 1802kW (2417shp) each
Maximum speed: 300km/h (190 mph)
Range: 800km (500 miles)
Service ceiling: 5000m (20,000ft)
Crew: 3
Armament: 2 x 12.7mm (0.5in)) M3M door guns, MU90 torpedoes, rockets and gun pods depending on operator

First customers
The first production TTH were delivered to the German Army in 2006, with the Italian and Australian armies receiving their first aircraft the following year. The Netherlands was the first country to receive the NFH in 2010, followed by Italy in 2011. Although deliveries had broadly gone to schedule there were a number of concerns from customers over issues such as the fragility of the composite floor and high levels of corrosion on the first NFH to conduct embarked operations. More worrying

RECONNAISSANCE AND UTILITY

was the temporary grounding of the Australian TTH fleet in 2010 due to an issue with the engines. Deliveries continued, however, and in 2012 Italy deployed five aircraft to Afghanistan for tactical transport and medical evacuation, Germany would deploy four aircraft there the following year.

Subsequently a high cabin version was developed for the Swedish armed forces with the first delivery taking place in 2007. This has a 24cm (9.4in) taller cabin, giving 1.82m (5.9in) of headroom, allowing it to be used as an air ambulance without contravening peacetime health and safety regulations. Nine aircraft are operated in the land role with

NHI NH90 TTH
ET807, together with the rest of the Spanish Army's NH90, is operated by Manoeuvre Helicopter Battalion 3 based at Agoncillo/Logroño.

a further nine configured for anti-submarine warfare. In February 2019 a maritime tactical transport (MTT) was announced, combining the folding rotors and strengthened undercarriage of the NFH with the transport-optimized fuselage of the TTH. Ten examples have been delivered to the Italian Navy to supplement their NFH fleet.

Spares and service issues
Parts availability and serviceability issues have unfortunately continued to dog the NH90 programme. Norway received the first of 14 NFH in 2011; by 2022, with one aircraft still to be delivered and the fleet averaging less than a fifth of the required flying hours, the Norwegian government terminated the NH90 contract and returned them to the manufacturer. Meanwhile, on the other side of the globe, Australia had been experiencing similar issues,

NH90 PT2
The second prototype, NH90 F-ZWTI was the first to test the fly-by-wire control system in 1997. Now retired, it has been restored by the Italian Army for display at the Volandia Flight Museum near Milan.

including a fleet grounding in June 2021 due to a lack of spares. Shortly after this Australia requested 40 UH-60M Black Hawks to replace its entire medium transport helicopter fleet, and by 2023 the NH90 had been retired, 15 years earlier than planned.

Although the early retirement of the Norwegian and Australian NH90 fleets has been a setback for the programme, the majority of the remaining 12 operators appear happy with the type. Indeed, the New Zealand Air Force has a serviceability rate of over 70 per cent and, despite only receiving its first example in 2011, were the first to have an aircraft reach 2000 flying hours.

RECONNAISSANCE AND UTILITY

Bölkow Bo 105

Developed as a light utility helicopter in the late 1960s, the Bo 105 was the first production light twin-engine helicopter and the first production helicopter to use a rigid rotorhead.

A fully articulated head features a hinge to allow each blade to flap up and down, and a second to allow them to move backwards and forwards in response to the changing aerodynamic loads as they rotate. In a rigid rotorhead these loads are instead accommodated by flexing of the blade root and head, which is mechanically simpler and more responsive. The greater stiffness of the rigid head also allows the Bo105 to perform aerobatic manoeuvres, such as loops, which are impossible for conventional helicopters.

Anti-tank helicopter

In 1975 the German Army acquired the first of 212 Bo 105 PAH-1 anti-tank helicopters that could be armed with six HOT wire-guided missiles. A further 100 Bo 105C were acquired in 1977 as observation aircraft. That year also saw the type become the first aircraft to be operated by the Royal Bahraini Air Force who received four.

IPTN in Indonesia produced 122 Bo 105 under licence, the last being delivered in 2008. These were predominantly for the Indonesian Armed Forces who used them in East Timor and the Papua conflict. South Korea similarly acquired a licence in the late 1990s with KAI building them for the Republic of Korea Army. These were fitted with electronic warfare, target acquisition and military communications systems to address a shortfall in reconnaissance capability.

Service today

Although now in limited service the Bo105 was developed into the Eurocopter EC135 and, via the BK117, the EC145. Both feature a hingeless rotorhead that retains the Bo 105's manoeuvrability.

Bo 105
Dimensions: Length: 11.86m (38ft 11in); Rotor diameter: 9.84m (32ft 3in); Height: 3m (9ft 10in)
Weight: 2500kg (5512lb) maximum take-off
Powerplant: 2 × Allison 250-C20B turboshaft engines, 310kW (420shp) each
Maximum speed: 242km/h (150 mph)
Range: 657km (408 miles)
Service ceiling: 5200m (17,000ft)
Crew: 1–2
Armament: SNEB rocket pod, HOT or TOW anti-tank missiles

MBB Bo-105M
Operated by the West German Army, the Bo-105M was a light transport and surveillance variant of the aircraft. 80+74 was retired and placed on the civil register in 2002 and is now located in the USA.

RECONNAISSANCE AND UTILITY

Bell OH-58 Kiowa

Originally developed for the light observation helicopter competition that was won by the Hughes OH-6, Bell's D-250 was an awkward-looking aircraft with limited internal capacity.

Seen over Croatia this is one of 16 ex-US Army OH-58D that were donated to the Croatian Air Force in 2016. Full operational capability was achieved in 2020 with the restoration of the ability to operate Hellfire missiles.

After losing to Hughes Bell redeveloped the aircraft for civil use as the Bell 206 with a streamlined fuselage, cargo bay and space for three passengers in the rear. With Hughes unable to meet the US Army's demand for observation helicopters, a further tender was issued in 1967, which Bell won and the 206 was designated OH-58A Kiowa in service use.

Powered by the same 190kW (258shp) Allison T63 turboshaft as the OH-6, the early OH-58 had a two-bladed, semi-ridged rotor system similar to that on the UH-1. Although not allowing the OH-58 the same degree of manoeuvrability as the OH-6, its high inertia provided superior auto-rotation characteristics. The view from the cockpit was excellent thanks to the curved main windscreen that has remained a feature of the type through all subsequent models, aside from a handful of OH-58C. Firepower was provided by optional 7.62mm (0.3in) M134 miniguns mounted to the fuselage immediately behind the cabin doors.

OH-58A deployed to Vietnam in August 1969 where it would operate much like the OH-6 in hunter-killer teams with AH-1 gunships and as an artillery observation post and battle damage assessment platform. Unlike later OH-58 models these had no external sensors and relied on the crews' eyes to detect enemy activity and assess the fall of gunfire.

While the OH-58B introduced minor improvements and was primarily for export, the next major version was the OH-58C, which was powered by a 310kW (415shp) version of the T63 and improved performance. Introduced in 1978, the C model brought a fully night vision goggles (NVG) compatible cockpit, an AN/APR-39 radar detection system, and on some aircraft the ability to carry AIM-92 Stinger air-to-air missiles. Around 275 OH-58A were converted to C models and the type was ultimately the last version to see service with the US Army, retiring in 2020.

Winning OH-58D

In 1981 the OH-58D was selected as the winner of the Army Helicopter Improvement Programme (AHIP), beating the Hughes OH-6-based proposal. AHIP was intended to fulfil a requirement for an improved scout

RECONNAISSANCE AND UTILITY

Bell OH-58D
10553 was operated by 4th Squadron of the 17th Cavalry during 1988's Operation Prime Chance, protecting shipping from attacks by Iran in the Persian Gulf. This was the first use of armed OH-58Ds, providing valuable information for future operations.

OH-58D

Dimensions: Length: 12.85m (42ft 2in); Rotor diameter: 10.67m (35ft); Height: 3.93m (12ft 10in)
Weight: 2495kg (5500lb) maximum take-off
Powerplant: 1 × Rolls-Royce T703 turboshaft engine, 485kW (650shp)
Maximum speed: 240km/h (149mph)
Range: 260km (161 miles)
Service ceiling: 4575m (15,000ft)
Crew: 2
Armament: Hellfire missiles, 70mm (2.75in) Hydra rockets, 12.7mm (0.5in) M3 cannon

helicopter, the most obvious alteration being the addition of a gyro-stabilized mast-mounted sight (MMS). This contained infrared and visual cameras and a laser rangefinder and target designator placed in a spherical housing above the main rotor. Performance was improved with the 484kW T703 engine, which drives a four-bladed main rotor. Constructed from composite materials, the new rotor and head were based on Bell's commercial work and offered increased agility with low vibration, the latter improving the reliability and performance of the MMS. Deliveries started in December 1985 and although initially intended to operate alongside the US Army's Apache gunships, the OH-58D was limited to the artillery observation role. This changed in 1988 when Kiowas from the 17th Cavalry were tasked with taking over from the OH-6 of the 160th SOAR in the Persian Gulf to protect merchant shipping from Iranian forces in Operation Prime Chance. These aircraft were modified for the role with armament and were unofficially designated AH-58D. Operating from two modified barges and primarily at night, these aircraft

RECONNAISSANCE AND UTILITY

were cleared to fire Stinger and Hellfire missiles, 70mm (2.75in) rockets, and 12.7mm (0.5in) cannon. Fifteen aircraft were modified in a process taking just under 100 days each.

Middle East action

Around 130 OH-58 deployed to the Middle East in support of Operations Desert Shield and Storm where they operated alongside AH-1 Cobras and AH-64 Apaches. The Kiowa force flew over 9000 hours in support of the operations while achieving 92 per cent mission capability. In January 1992, after Desert Storm had finished, the US Army contracted for the first batch of OH-58D to be upgraded to Kiowa Warrior standard. This included the armament systems integrated to the Prime Chance aircraft along with avionics, structural and transmission upgrades. Kiowa Warriors saw extensive service in Iraq and Afghanistan, where they frequently operated in direct support of the infantry rather than a scouting role. In the latter theatre operations were regularly conducted in the mountains, operating as high as 4250m (14,000ft) with occasional enemy fire coming from above the aircraft.

Despite its utility and heavy utilization during its decades of service, the US Army cancelled all plans for upgrades or replacement of the OH-58D in early 2014 as part of a restructuring. This ultimately saw its role taken over by a combination of the more-expensive-to-operate AH-64 and UAVs. While all US OH-58D were retired by 2017 and the C model a few years later, a number of other countries continue to operate both models including Croatia, Greece, Taiwan and, somewhat ironically, Iraq.

Bell OH-58B Kiowa

Australia was an early operator of the A model Kiowa, leasing eight aircraft for use in Vietnam in 1971. Subsequently 56 would be purchased, the majority of which were licence built by Commonwealth Aircraft Corporation in Australia. The last of these were retired in 2018.

Tunisia acquired 24 surplus OH-58D with the first deliveries occurring in January 2017. This example is seen taking part in Tunisia's first International Air and Defence Exhibition (IADE) in 2020 at Djerba International Airport.

OH-58B

Dimensions: Length: 12.85m (42ft 2in); Rotor diameter: 10.67m (35ft); Height: 3.93m (12ft 10in)
Weight: 1451kg (3198lb) maximum take-off
Powerplant: 1 × Allison T-63-A-720 turboshaft engine, 313kW (420shp)
Maximum speed: 220km/h (136mph)
Range: 490km (304 miles)
Service ceiling: 4575m (15,000ft)
Crew: 2
Armament: 1 x 7.62mm (0.308in) M134 minigun

RECONNAISSANCE AND UTILITY

Bell CH-146 Griffon

Based on the Bell 412 Twin Huey, the CH-146 Griffon has been modified to fulfil the requirements of the Canadian Armed Forces replacing their existing UH-1H and UH-1N aircraft.

A hundred Griffon were acquired in SAR and utility configurations, with the latter intended to carry a section of eight troops. Based on a civilian aircraft the CH-146 can be optionally configured with extended range fuel tanks, armour, door guns and other specialized equipment depending on the intended role.

The Griffon entered service in 1995 and was employed the following year on operations in Haiti where it would return again in 2004 after the coup d'état, and again in 2010 to assist in aid efforts after the country was struck by an earthquake. At the turn of the century CH-146 were deployed to Bosnia in support of Canada's contribution to the NATO stabilization force. They have also seen widespread use across Canada providing aid to the civil authorities during natural disasters. In 2008 eight aircraft were modified for operations in Afghanistan; fitted with the 7.62mm (0.3in) M134 minigun, they were intended to act as escorts for Canadian CH-147 Chinooks.

Bell CH-146 Griffon
The second to last Griffon to be taken on strength by the Canadian Armed Force on the 17th of December 1997, 46499 is painted in the standard scheme for those aircraft used for tactical aviation.

These operations were hampered by the weight of additional equipment required for the theatre and the aircraft's reduced performance at high altitudes. Although this is a problem for all helicopters, the CH-146 was at the limits of its capabilities in Afghanistan and a fatal accident in 2009 was partially attributed to the aircraft's poor performance.

From 2024 the CH-146 are being upgraded with new avionics, sensors and PT6T-9 engines with an electronic control system. Eighty-two aircraft in the active fleet are being upgraded and are expected to remain in service until the mid-2030s, operating from 10 bases across Canada.

CH-146
Dimensions: Length: 17.09m (56ft 1in); Rotor diameter: 14.02m (46ft); Height: 4.60m (15ft 1in)
Weight: 5398kg (11,900lb) maximum take-off
Powerplant: 2 × Pratt & Whitney PT6 turboshaft engines, 960kW (1290shp) each
Maximum speed: 260km/h (162mph)
Range: 656km (407 miles)
Service ceiling: 6100m (20,000ft)
Crew: 3
Armament: 7.62mm (0.3in) M134 minigun, 12.7mm (0.5in) GAU-21 cannon

RECONNAISSANCE AND UTILITY

Eurocopter AS365 Dauphin II

Developed from the single engine Aérospatiale Dauphin, the AS365 Dauphin II first flew in 1975 and became a commercial and military success with a range of variants and licence production in Korea and China.

The Dauphin II is a twin-engine helicopter with a four-bladed main rotor and a Fenestron tail rotor; the streamlined fuselage has seating for a crew of two and up to 12 passengers depending on the configuration. Unlike the earlier aircraft the AS365 uses a Starflex rotorhead, designed by Aérospatiale. This replaces mechanical hinges with a flexible composite rotorhead and elastomeric bearings. This provides greater responsiveness than a hinged rotorhead without the vibration levels and cost associated with a rigid rotor system. Although early aircraft had a fixed undercarriage the introduction of the AS365N in 1979 brought retractable gear, which together with the streamlined design helped the Dauphin II gain three point-to-point speed records between Paris and London the following year.

In 1979 the Dauphin II was selected by the United States Coast Guard for the short-range search and rescue role and entered service in 1985. Designated the HH-65A Dolphin it was the first helicopter to incorporate a four-axis autopilot, which permits hands-off searches to be flown allowing the pilot to concentrate on other tasks as well as automatically bringing it down to a 15m (50ft) hover over a designated spot. Initially equipped with Lycoming LTS101 engines, to meet American content requirements these were ultimately replaced with the Turbomeca Arriel used by other operators due to reliability issues.

Panther introduced

A dedicated military version of the Dauphin was introduced in 1986 as the AS365M, later AS565, or Panther and is based on the AS365N. It can be configured for utility, transport, anti-tank and maritime operations. While a variety of cockpit layouts are available the vast majority are equipped with a glass cockpit, something the Dauphin II pioneered on helicopters, 4-axis autopilot and digital engine management system to minimize the pilot workload.

Eurocopter AS565 Panther
Delivered in 1997, Panther 505 is operated by the French 36th Airborne Command and Control Wing (36th CCW). It was reportedly the first to be updated to 'Standard 2' configuration, which gained a Link 11 tactical data link, upgraded avionics, and radios.

AS565 Panther
Dimensions: Length: 13.68m (44ft 11in); Rotor diameter: 11.94m (39ft 2in); Height: 3.97m (13ft)
Weight: 4300kg (9480lb) maximum take-off
Powerplant: 2 × Turbomeca Arriel 2C turboshaft engines, 635kW (852shp) each
Maximum speed: 306km/h (190mph)
Range: 827km (514 miles)
Service ceiling: 5865m (19,242ft)
Crew: 1–2
Armament: 20mm (0.78in) M621 cannon pods, AS.15 and HOT air-to-surface missiles, Mistral air-to-air missiles, 68mm (2.6in) and 70mm (2.75in) rocket pods, lightweight torpedoes

The Brazilian Army were an early adopter of the type with 36 Panthers being locally produced by Helibras. Used in the light transport role, these aircraft went through a mid-life upgrade in the 2010s to AS365K2 standard, which introduced new avionics, weather radar and Arriel 2C2CG engines delivering 40% per cent more power.

Naval versions of the Panther can be configured with the L3Harris HELRAS dipping sonar and magnetic anomaly detector for anti-submarine warfare, while a range of radars and passive sensors can be fitted for the anti-surface warfare role. Armament options include anti-ship missiles, torpedoes and cannon pods. For shipboard operations the Panther can be fitted with a deck lock on the bottom of the fuselage, which attaches to a grid in the host ship's flight deck securing it in rough seas until lashings can be fitted. An emergency flotation system is also fitted to prevent the aircraft sinking long enough for the crew to escape a ditched aircraft. The French Navy operates a Panther squadron, 35F, primarily providing individual aircraft to operate from escort vessels, while 34F uses Dauphins for SAR operations including a detachment on the carrier Charles de Gaulle. French Navy Panthers have been used for counter piracy operations in the Gulf of Aden, operating from frigates and in 2024 one reportedly downed a Houthi drone with gunfire. From 2009 the 15 aircraft fleet was upgraded with new avionics, a data-link and a defensive aids suite based on the Eurocopter Tiger's. Other naval operators include Saudi Arabia, whose aircraft fired TT.15 missiles at Iraqi patrol boats in the 1991 Gulf War, Mexico and Israel. Indonesia meanwhile has locally produced the type for the ASW role.

Upgrade to EC155

First flying in 1997 the EC155 is an upgraded Dauphin II with 30 per cent more cabin space and 130 per cent more baggage capacity. Although externally very similar to the earlier aircraft, it has a five-bladed main rotor and more powerful 697kW (935shp) Arriel 2C2 engines. While the Dominican Republic's Air Force is the only military operator of the EC155, South Korea is developing the type under licence as a light armed helicopter (LAH) to replace its MD500 scouts and AH-1S gunships. At the same time Korea's KAI became the sole production line for the civil variant of the EC155 from 2018. The LAH is distinct from other military variants of the Dauphin family in having a nose-mounted 20mm (0.78in) M197 Gatling cannon, above which is mounted an electro-optical sensor turret. Fuselage-mounted stub wings to the rear of the cabin doors can carry up to four Chun-gum multi-mode anti-armour missiles, or 70mm (2.75in) Hydra rockets in pods. Initial production deliveries of the LAH were due to start in late 2024.

Transfer of the EC155 production line to Korea has freed up capacity to begin assembly of its replacement, the H160, a more advanced twin-engine helicopter with a similar basic configuration to the Dauphin. The H160M Guépard is the military version and 174 have been ordered by France to replace five different types including the Dauphin and Panther.

Eurocopter AS365 N3 Dauphin
Operated in a pseudo civilian scheme, although retaining its military serial, this is one of five Dauphin AH.1 operated by 658 Squadron Army Air Corps in support of 22 Regiment of the SAS. It is based at their headquarters in Stirling Lines.

RECONNAISSANCE AND UTILITY

RECONNAISSANCE AND UTILITY

Aérospatiale Gazelle

Initially developed by Sud Aviation, which became part of Aérospatiale in 1970, the Gazelle was designed to meet a French Army requirement for an Alouette III replacement in the light observation role.

Although it used the Turbomeca Astazou turboshaft engine, the Gazelle was significantly different from the earlier aircraft. The three-bladed main rotorhead is semi-rigid and constructed from composite material as are the rotor blades, and at the time of its introduction was considered low maintenance with no requirement to lubricate the bearings. More obvious, although in some ways easier to develop, was the Fenestron tail rotor that was introduced on the second prototype Gazelle. Housing the 13-bladed rotor in a duct in the vertical stabilizer protects it from striking obstructions when operating in confined areas and reduces the acoustic signature. Although greater power is required to maintain heading in slow speed flight at higher speeds, thanks to the large vertical fin, less power is required than with a traditional rotor system.

Easy to operate the Gazelle has good high-speed performance and is stable enough that some variants do not feature an auto-stabilization system, reducing weight and complexity. With the addition of a Honeywell Duplex autopilot, the Gazelle was the first helicopter to be certified for single pilot operations in instrument

Westland SA-341B Gazelle AH.1
XX457 was operated by the Advanced Rotary Wing Flight at Middle Wallop for flying training, hence the day-glow panels. Damaged during take-off in 1987 it became a ground instructional airframe.

RECONNAISSANCE AND UTILITY

flying conditions. The aircraft can seat five people, allowing for four passengers if being operated in a single pilot configuration.

French and British service
The Gazelle entered service with the Aviation Légère de l'Armée de Terre (ALAT, French Army Light Aviation group) in 1971 who initially acquired 166 SA341F powered by the Astazou IIIC. These fulfilled a light utility role but would be followed by the SA341M armed with four HOT missiles for anti-tank operations. Subsequently the SA341F2 would be introduced with a 20mm (0.78in) M621 cannon mounted on the starboard side and cabin-mounted magazine for up to 300 rounds. The F2 is used in the fire suppression role. The SA341M Gazelle ATAM (air-to-air missile) were modified to fire Mistral air-to-air missiles, with an effective 6km (3.7 mile) range.

Initially introduced for the Gulf War and able to carry two missiles, the SA341ML1 was upgraded to carry four with a T200 sight. The gyro-stabilized sighting system for the missiles is mounted on top of the fuselage above the left-hand pilot's seat and can also be used for observation and reconnaissance. As part of a cross-licensing deal the UK's Westland Helicopters would produce 262 Gazelles for the British armed forces, with Aérospatiale having a similar arrangement for Lynx destined for the French Navy.

The British Army accepted the first of 158 SA341B in 1974. These would be used for various operational and training roles with the Army Air Corps and the Royal Marines, primarily used as scouts and rarely armed. At the same time the Royal Navy and Royal Air Force would acquire Gazelles for pilot training, making it one of the few types to have been operated by all four services.

The Gazelle's utility as an anti-tank aircraft was demonstrated during the 1982 Lebanon War where Syria used them to slow the advance of Israeli tanks using HOT missiles. Although only around seven tanks were actually destroyed the psychological effect was not insignificant and more aircraft would be acquired after the war. Iraq meanwhile would team its Gazelles with Mil Mi-24 Hinds to counter Iranian attacks during the eight year-long Iran–Iraq war. France deployed around 50 SA341M as part of its contribution to Operation Desert Storm in 1990–91; during the advance into Kuwait they were employed attacking Iraqi vehicles and in a counter air role. British Gazelles were also active during the conflict operating as scouts for the Army Air Corps' Lynx anti-tank helicopters.

Westland SA-341B Gazelle AH.1
The Gazelle was operated in several schemes during its time with the Army Air Corps. XZ304 is seen in the one most commonly used in its final years of service and was operated by 6 Flight AAC based at RAF Shawbury, until it disbanded in 2009.

Aérospatiale Gazelle
Dimensions: Length: 13.68m (44ft 11in); Rotor diameter: 11.94m (39ft 2in); Height: 3.97m (13ft)
Weight: 4300kg (9480lb) maximum take-off
Powerplant: 2 × Turbomeca Arriel 2C turboshaft engines, 635kW (852shp) each
Maximum speed: 306km/h (190 mph)
Range: 827km (514 miles)
Service ceiling: 5865m (19,242ft)
Crew: 1–2
Armament: 20mm (0.78in) M621 cannon pods, AS.15 and HOT air-to-surface missiles, Mistral air-to-air missiles, 68mm (2.6in) and 70mm (2.75in) rocket pods, lightweight torpedoes

RECONNAISSANCE AND UTILITY

An SA 342 Gazelle utility helicopter with French forces in Djibouti makes an aerial pass during a ground and air display near Arta Range, Djibouti, in January 2022.

SOKO of Yugoslavia and Egypt's Arab British Helicopter Company (ABHCO) both had licences for local production of the Gazelle.

The former began production in the early 1970s; consequently when the civil wars that would ultimately lead to the dissolution of Yugoslavia broke out in 1991, they were in use by all sides. French and British Gazelles also saw service in the region as part of the UN intervention to end the civil wars that endured throughout the decade.

Updates for the 21st century
While Gazelles saw operational service during the early 2000s, including Iraq, Afghanistan and counter piracy off the Horn of Africa there was a requirement to update them for the 21st century. ALAT's aircraft gained the Viviane thermal imaging system, improving its night reconnaissance and attack capabilities. This replaced the original weapons sight with a new unit housing television and thermal cameras as well as a laser rangefinder and target designator. With multiple zoom levels the system allowed targets to be detected, marked and engaged up to a range of 5km (3 miles).

In the UK where the aircraft was primarily used for operations in Northern Ireland from the early 2010s, an upgrade added a Garmin GNT650H GPS, 8.33KHz VHF radios, GTS800 TAS (traffic advisory system) and a SN3500 electronic navigation display. The new cockpit is fully NVG compatible while the GPS was the first on a UK military aircraft to allow GPS instrument approaches to be flown.

Although the Gazelle left UK service in 2023 and is being phased out by France to be replaced by the H160M, it remains in service with around 20 other operators.

Bell 412

While the UH-1N Twin Huey was a success, and continues in widespread use, in the late 1970s Bell worked on a new version that would feature a four-bladed main rotor.

The original Huey and Twin Huey used a robust twin-bladed rotor that had good lifting performance but suffered from vibration and a limited top speed. Moving to a four-bladed system reduces vibration and noise levels and allows for a 20–25km/h (12–15mph) increase in maximum and cruise speeds. At the same time the rotor diameter could be reduced by around 0.6m (2ft) while achieving a better maximum all-up mass.

While the Twin Huey has the Bell model number 212, adoption of a four-bladed rotor led to the new type being designated the 412. The first two aircraft were in fact modified from 212s and fitted with a new hingeless main rotorhead. The hub is composed of two titanium flex-beam yokes stacked on top of each other and offset 90°. Steel spindles are used to attach the new composite rotor blades, with elastomeric bearings and dampers used to reduce maintenance requirements. Mechanical pendulum dampers mounted on the leading and trailing edges of each yoke are designed to absorb vibrations. The 412 retains the two-bladed tail rotor of the 212.

First production

First deliveries of the Bell 412 took place in 1981, the same year Augusta began licence production in Italy. Bahrain, Venezuela and Guatemala were among the countries to order a small number for their militaries in the early 1980s. In 1986 Indonesia's IPTN delivered the first of over 100 licence-built aircraft many going to the Indonesian Army, which in 2024 had a fleet of over 60 with new examples still being acquired. The Italian Army meanwhile took delivery of Augusta-built aircraft from the mid-1980s, many of which remain in service in 2024 alongside locally built Bell 212s. Canada acquired 100 locally built CH-146 from the mid-1990s, covered separately.

Bell 412EPI
Dimensions: Length: 17.09m (56ft 1in); Rotor diameter: 14.02m (46ft); Height: 4.57m (15ft)
Weight: 5398kg (11,900lb) maximum take-off
Powerplant: 2 × Pratt & Whitney PT6 turboshaft engines, 960kW (1290shp) each
Maximum speed: 260km/h (160 mph)
Range: 980km (609 miles)
Service ceiling: 6100m (20,000ft)
Crew: 1–2
Armament: 7.62mm (0.3in) M134 minigun, 12.7mm (0.5in) GAU-21 cannon or similar depending on the operator

---

Bell 412EP
The Chilean Air Force operate 15 Bell 412EP, alongside eight UH-1H models, all of which are used in the utility role.

RECONNAISSANCE AND UTILITY

Updates

In response to changing demands and improving technology Bell continued to update the basic airframe. The 412EP, Enhanced Performance, is powered by the P&W PT6T-3DF and was selected by the RAF as its advanced rotary wing training aircraft from 1997 until 2018, and for search and rescue duties in Cyprus until 2023.

Bell 412HP

The Royal Norwegian Air Force operates a mix of ten SP and eight HP Bell 412s. 165 is an HP model which was deployed to Kosovo in the mid-2000s for duties with the United Nations. Usually it is based in Bardufoss, Norway, supporting the army and special forces.

The 412EPI features a glass cockpit and the PT6T-9 version of the engine, which has an electronic control system. The EPI introduced the BLR Aerospace fast fin system to production aircraft. Also available as a retrofit option, this reprofiles the vertical stabilizer to minimize interference with the thrust from the tail rotor. Additionally, strakes are added to the left-hand side of the tail boom, which breaks up downwash from the main rotor on that side, allowing the boom to act as an aerofoil, providing a counter-torque force. As well as improving the maximum all up mass it also improves heading control, allowing sideways flight at 83km/h (52mph) without issue – 20km/h

Bell Griffin HAR Mk2

Based in Cyprus the RAF's 84 squadron provides SAR coverage to the UK's base areas. ZJ703 was one of four Griffin HAR Mk2 it operated between 2003 and 2023 when it was replaced by the Westland Puma HC Mk2.

(12.4mph) faster than the 412EP. Despite having been in production for over 40 years the Bell 412 continues to be a popular choice. In 2020 the Japanese Ground Self Defence Force (JGSDF) contracted with Subaru for the first of 150 UH-2. Based on the Bell 412EPI these are planned to replace the JGSDF's fleet of UH-1J Hueys, and deliveries began in 2022.

RECONNAISSANCE AND UTILITY

Changhe Z-11WB

At times claimed as the first domestically designed Chinese helicopter, the Changhe Z-11 was based on the Eurocopter AS350 Squirrel that first flew in 1974 and is in widespread use as a utility helicopter.

Work began on the Z-11 in 1989, potentially by reverse engineering a Squirrel acquired in the USA. First flight followed in December 1996 and civil certification was granted in 2001. The Z-11's conventional layout has a three-bladed main rotor and two-bladed tail rotor powered by a single 510kW (680shp) Liming WZ-8 turboshaft engine, based on the Turbomeca Arriel.

A skid undercarriage is used while, unlike the Squirrel, the early models had a dolphin nose added to the fuselage. The first military version was the Z-11W, which features a sensor turret above the fuselage and provision for a gun or rocket pods and missiles on a braced beam aft of the rear doors. Alternatively, cabin-mounted door guns can be fitted.

The Z-11WB, also known as 'Kuang' (Buzzard), first flew in 2015. The most significant change is the forward fuselage, which has a rounded nose similar to the HC120 with the sensor turret placed beneath it when fitted. Some aircraft have also been seen with a radome fitted to the front of the fuselage. In service with the People's Liberation Army, the Z-11WB is used as a reconnaissance and attack aircraft similar to the US Army's MH-6, including the ability to carry a small number of troops. In addition to conventional weaponry, it can also carry the SW-6 UA; this 20kg (44lb) vehicle can carry a 5kg (11lb) payload and, once launched, can operate independently for around one hour.

Changhe Z-11WB

Seen at the China International Aviation and Aerospace Exhibition in Zhuhai, Guangdong province, in November 2016, this Z-11WB carries the Aviation Industry Corporation of China logo on the tail and abbreviated titles in Chinese characters on the boom. It is shown with a quartet of demonstration TL-2 air-to-surface missiles mounted on the port pylon.

Z-11WB

Dimensions: Length: 11.24m (36ft 11in); Rotor diameter: 10.69m (46ft); Height: 3.14m (10ft 4in)
Weight: 2250kg (4960lb) maximum take-off
Powerplant: 1 × Liming WZ-8 turboshaft engine, 510kW (680shp)
Maximum speed: 260km/h (160 mph)
Range: 661km (411 miles)
Service ceiling: 5270m (17,290ft)
Crew: 1–2
Armament: 7.62mm (0.3in) CS/LM12 door-mounted minigun, 23mm 0.9in) gunpod, 57mm (2.2in) rockets, HJ-8 and HJ-10 air-to-surface missiles, TY-90 air-to-air missiles, SW-6 UAV

RECONNAISSANCE AND UTILITY

Bell UH-1Y Venom

Having operated the UH-1N Twin Huey since the 1970s, in 1996 the United States Marine Corps (USMC) launched a programme to replace it alongside the AH-1W.

Although this resulted in an aircraft with a four-bladed main rotor, it differed significantly from the Bell 412 development of the UH-1N. The new main rotor has four composite main rotors, two of the blades folding to allow the 'Yankee' to fit in the same ground footprint as the UH-1N, vital for shipboard deployments. The four-bladed main rotor is on the left-hand side of the vertical stabilizer, the opposite site to the earlier aircraft. Power is provided by two 1153kW (1546shp) GE T700 engines connected to a new transmission system. This dynamic system is shared with the AH-1W simplifying the logistics and maintenance situation when deploying onboard one of the US Navy's amphibious warfare ships. The new rotors and engines give the UH-1Y a maximum take-off weight almost three tonnes heavier than the Bell 412 and four more than the UH-1N. At the same time, it makes the UH-1Y quieter, faster and more manoeuvrable than earlier Huey models.

To take advantage of the extra lift capacity the fuselage has been extended with a 0.53m (20in) plug between the cockpit and cabin. Meanwhile, the new aircraft has consolidated upgrades that were made to the UH-1N and introduced modifications that had been deferred to save money. The NVG compatible cockpit features an integrated avionics system with four multifunction displays presenting information from the mission, weapons, communications and navigation systems. Mounted under the nose is a BRITE Star thermal imaging and laser designation system. Self-defence is provided by an APR-39 radar warning receiver and AAR-47 missile and laser detection system.

Bell UH-1Y
Dimensions: Length: 17.78m (58ft 4in); Rotor diameter: 14.88m (48ft 10in); Height: 4.45m (14ft 7in)
Weight: 8391kg (11,840lb) maximum take-off
Powerplant: 2 × General Electric T700-GE-401C turboshaft, 1153kW (1546shp) each
Maximum speed: 304km/h (189 mph)
Range: 600km (373 miles)
Service ceiling: 6100m (20,000ft)
Crew: 2
Armament: 7.62 M240D or 12.7mm (0.5in) GAU-21 machine guns, 70mm (2.75) Hydra 70 or APKWS rockets

Bell UH-1Y Venom
0490 was the first UH-1Y Venom delivered to the Czech Air Force in August 2023. The aircraft are operated alongside the CzAF's AH-1Z Vipers by the 221st Helicopter Squadron at Náměšt nad Oslavou Air Base in the east of the country.

RECONNAISSANCE AND UTILITY

Chaff and flares are dispensed from ALE-47 dispensers mounted on either side of the forward fuselage and tail boom. The greater mass has, however, resulted in a smaller ground to fuselage clearance and a higher nose-up hover attitude of 8°.

Armament

The UH-1Y can be armed with 19 round launchers for 70mm (2.75in) Hydra rockets. These can be unguided or fitted with the advanced precision kill weapon system (APKWS) laser guidance unit. This gives the rockets a circular error of probability (CEP) of less than 0.5m (1.6ft), meaning half of the rounds fired will be within that distance of the aiming point. Door mounts are also provided for 7.62mm (0.3in) or 12.7mm (0.5in) machine guns for close protection.

Although the US Marine Corps (USMC) initially remanufactured UH-1N airframes into UH-1Y the decision was taken in 2005 to acquire new build aircraft. This had a marginal cost difference but crucially avoided a drop

Bell UH-1Y Venom
This UH-1Y is operated by Marine Light Attack Helicopter Squadron 169 (HMLA-169), based in Camp Pendleton, California. The squadron currently operates 12 UH-1Y alongside 15 AH-1Z Vipers, and has made regular deployments to Afghanistan and embarked as part of a Marine Expeditionary Unit.

in the number of aircraft available to be deployed in support of Operation Iraqi Freedom. Production deliveries commenced in 2007 with the first operational deployment taking place in 2009 aboard the USS Boxer. UH-1Y also deployed to Afghanistan in 2012 when aircraft of HMLA-469 were involved in the defence of Camp Bastion after Taliban fighters infiltrated the base. Together with an AH-1W from the squadron, two of the Hueys armed with door guns took off under fire and provided overwatch to troops on the ground as they neutralized the attackers.

The USMC received 160 UH-1Y with production completed by 2018, however, it was restarted in the early 2020s to fulfil an order for eight aircraft from the Czech Air Force. These aircraft, the first of which was delivered in 2023, will be supplemented by a further two from USMC stocks in recognition of Czech support for Ukraine.

RECONNAISSANCE AND UTILITY

Kamov Ka-226

A light utility helicopter, the Ka-226 uses Kamov's co-axial contra-rotating rotor system, which allows all the available power to be used for lift.

Essentially a twin turboshaft development of the earlier twin piston Ka-26 and single turboshaft Ka-126, all three types share the same NATO designation of Hoodlum. To enable a variety of roles to be fulfilled, the Ka-226 carries mission pods in place of a conventional fuselage. These are fitted behind the fixed cockpit section and can be configured for passenger carriage, medical evacuation, cargo or even firefighting duties. Modules can be swapped in around two hours; alternatively it can be operated without any module fitted as a flying crane able to lift up to 1500kg (3306lb).

The mission modules have led to the use of a high twin-boom tail with the engines exhausting into the gap between them, which enables the modules to be removed from the rear with minimal interference. As with all co-axial helicopters, large vertical stabilizers with rudders are fitted to provide directional control at higher speeds.

Unusually for a Russian helicopter it does not use domestically produced engines. Early aircraft were powered by the 336kW (450shp) Rolls-Royce/Allison 250-C20R/2 while the Ka-226T uses French engines, two 435kW (580shp) Turbomeca Arrius 2G1. The cockpit meanwhile features a modern avionics suite with NVG compatibility. First flying in 1997, the Ka-226 was ordered by the Russian Aerospace Forces and the Ministry of Internal Affairs in 2011 with deliveries from 2015. The Ukrainian Navy operates a single example that was originally acquired by the State Emergency Service in 2008 but not put into service. Syria is reported to have had two examples both of which were destroyed by Israel in late 2024 in an effort to prevent them falling into the hands of hostile forces.

Kamov Ka-226T
This Ka-226T is carrying the passenger module, the break line with the fuselage can be seen forward and above the cabin door. It is wearing the standard camouflage scheme of the Russian Aerospace Forces currently the major operator of the type.

Changhe Ka-226
Dimensions: Length: 8.1m (25ft 7in); Rotor diameter: 13m (42ft 8in); Height: 4.15m (13ft 7in)
Weight: 3800kg (8379lb) maximum take-off
Powerplant: 2 × Turbomeca Arrius 2G1 engines, 435kW (580shp) each
Maximum speed: 250km/h (155 mph)
Range: 600km (373 miles)
Service ceiling: 6200m (20,300ft)
Crew: 1–2
Armament: None

RECONNAISSANCE AND UTILITY

Harbin Z-9

A licence-built version of the Aérospatiale Dauphin II, the Harbin Z-9 has been developed into a distinct line of aircraft

The Z-9A first flew in 1981 and was constructed from parts supplied by France. It was followed by the Z-9B in 1992 with around 72% Chinese-manufactured components including the Liming WZ-8 engines.

Initially used as a light utility aircraft, subsequent models were armed with machine guns and rocket pods. From 2010 the improved Z-9B was delivered with an enlarged nose housing a weather radar for search and rescue. These aircraft can also be fitted with a hoist, searchlight and infrared camera. The Z-9B has been exported as the Z-9EA and Z-9EH to countries in Africa and South-east Asia. The Z-9WA developed in 2000 is a night attack version analogous to the Eurocopter Panther. The nose has been reconfigured to mount an electro-optical turret on the underside, stub wings allow the carriage of up to eight anti-tank or air-to-air missiles and a radar warning receiver is fitted. The NVG compatible cockpit is also protected by armour.

Harbin Z-9
Wearing the standard colour scheme for the People's Liberation Army Navy (PLAN), Z9-0046 was operating from the *Zhoushan* in 2009 when she transferred officers to HMS *Cornwall* to discuss counter piracy operations in the Gulf of Aden.

The Z-9C is the anti-submarine warfare version of the aircraft and was introduced to the People's Liberation Army Navy (PLAN) in the 1990s. It is fitted with a Type 605 dipping sonar and can also carry two homing torpedoes. An enlarged nose houses a radar while a searchlight, electro-optical turret and hoist can be fitted as role equipment. The Z-9D meanwhile is configured for the anti-shipping role with the ability to carry up to four TL-10B anti-ship missiles able to neutralize fast attack craft and small ships. The recently developed Z-9F appears to combine both roles with a new surface search radar, dipping sonar and glass cockpit. Naval variants have been exported to Pakistan for use on their *Zulfiquar* frigates.

Harbin Z-9B
Dimensions: Length: 12.11m (39ft 9in); Rotor diameter: 11.94m (39ft 2in); Height: 4.01m (13ft 2in)
Weight: 4100kg (9039lb) maximum take-off
Powerplant: 2 × Liming WZ-8A turboshaft, 632kW (848shp) each
Maximum speed: 305km/h (190 mph)
Range: 1000km (620 miles) with auxiliary fuel tank
Service ceiling: 4500m (14,800ft)
Crew: 2
Armament: 23mm (0.9in) AM-23 cannon, HJ-8 anti-tank or TY-90 air-to-air missiles, Yu-7 or Yu-11 torpedo

RECONNAISSANCE AND UTILITY

Harbin Z-20

Although the People's Liberation Army Air Force (PLAAF) acquired 24 Sikorsky S-70C Black Hawk aircraft in the early 1980s, due to the evolving political situation between the USA and China further orders were not possible, leading to the development of the Z-20 programme.

Clearly based on the Black Hawk, the Z-20 is not a direct copy of the US aircraft. An immediate difference is the use of a five-bladed main rotor versus the four-bladed unit on the older aircraft. These are driven by two Chinese-developed 1600kW (2100shp) WZ-10 engines while the aircraft also features active vibration reduction systems and fly-by-wire controls, the latter still rare in helicopters. The cockpit features five multifunction displays while a night vision system is fitted under the nose with a forward-facing radar mounted above it.

In addition to ECM antenna and laser warning receivers, a satellite communications antenna is mounted in a distinct dome on top of the tail boom. Defensive chaff and flare dispensers are also mounted around the aircraft. Assembly of the first airframes began in 2010 and the first flight took place in late 2013. Series production is believed to have begun in 2018 with deliveries to the People's Liberation Army Ground Force (PLAGF) beginning shortly after. These aircraft are typically seen in an all-over dark green, almost black, colour scheme. Although early aircraft had sideways exiting engine exhausts, since early 2021 examples have been seen with them directing the hot gases upwards into the rotor downwash to reduce the infrared signature.

Z-20T

Believed to have entered service in early 2024, the Z-20T is an armed variant fitted with stub wings able to carry KD-9 anti-tank missiles and rocket pods. It also features an electro-optical (EO) turret above the nose replacing the radar. While fulfilling a similar role to the US MH-60L, there are indications China is developing an air-to-air refuelling capability for the Z-20T.

The PLAAF operate the Z-20KA version, generally seen in an all-over light grey scheme, as a transport aircraft. It also uses the Z-20KS SAR model, which has an electro-optical turret mounted in front of the right main undercarriage and a steerable searchlight in front of the left. The PLAAF Z-20 that have been seen operating with People's Liberation Army Navy (PLAN) ships have a different tailplane, like that on the PLAN's Z-20F, and an unfaired tailwheel, and may be a new sub-type.

Harbin Z-20

Clearing showing its similarities with the Sikorsky Black Hawk, LH982281 was one of the earlier production aircraft and took part in the flying display at the 2022 Zhuhai Air Show.

Harbin Z-20
Dimensions: Length: 20m (65ft 7in); Rotor diameter: 16m (52ft 6in); Height: 5.3m (17ft 5in)
Weight: 10,000kg (22,046lb) maximum take-off
Powerplant: 2 × WZ-10 turboshaft, 1600kW (2100shp) each
Maximum speed: 360km/h (220 mph)
Range: 560km (350 miles)
Service ceiling: 6000m (20,000ft)
Crew: 2
Armament: Machine guns and cannon, AKD-9 and AKD-10 air-to-surface missiles, TY-90 air-to-air missile

RECONNAISSANCE AND UTILITY

Naval variants

Naval variants of the Z-20 were developed in the late 2010s and are believed to have entered service by 2024. The Z-20F is an ASW variant, similar to the MH-60R Seahawk. For shipboard operations it has a folding main rotorhead and tail with the horizontal stabilizer also folding to minimize the deck space taken up. As with the Seahawk, the tail wheel has been moved forwards to a position under the fuselage/tail boom joint, which allows landings on the smaller flight decks found on frigates and destroyers. An orange dome is mounted on the underside of the tail on all naval versions, and on the PLAAF sub-type noted above. This is believed to be a position marker that breaks away in the event the aircraft ditches; it may also house the flight data recorder as on some Western types. A 360° search radar is mounted under the nose with an EO turret above it while a dipping sonar is lowered from an opening in the bottom of the fuselage. The left-hand cabin door has been replaced with a 25-round sonobouy launcher while Yu-11 torpedoes can be carried on two stub pylons either side of the rear fuselage. Stub wings above the main gear can be used for YJ-9 air-to-surface missiles. The Z-20J is a naval transport and assault aircraft and differs from the F model in not having a search radar or ASW equipment while the EO turret is placed below the nose. It does, however, retain the folding rotors and tail section, and repositioned tailwheel of the J and is finished in the same all-over light grey colour scheme. It also retains the forward stub wings for missile carriage.

The Z-21

While the Z-20 has been developed to perform a similar range of roles to the Black Hawk family, 2024 saw the first images of the Z-21. This takes the transmission system and tail of the Z-20 and pairs it with an Apache-style tandem cockpit to create a heavyweight gunship to compliment the Changhe Z-10. While images seen to date are of poor quality, an under-fuselage cannon is clearly visible along with stub wings with a total of six hardpoints. An EO turret is placed on the nose while cheek fairings are believed to hold avionics, again as on the AH-64. As per the latest Z-20 the exhausts vent upwards into the rotor wash. Some examples have also been seen with what is believed to be a millimetre band radar mounted above the main rotor.

Although the Z-20 is the result of reverse engineering, this has allowed China to develop an array of aircraft able to fulfil different roles in a relatively short period of time. With the Z-21 it has also gone beyond the capabilities of the Black Hawk on which it is based.

RECONNAISSANCE AND UTILITY

Eurocopter AS532 Cougar

Since 1990 Eurocopter has marketed military derivatives of the Super Puma family as the AS532 Cougar, or latterly as the H215M, now production is under the Airbus umbrella. Aircraft built prior to 1990 by Aérospatiale used the AS332 Super Puma designation.

Based on the earlier Sud Aviation SA330 Puma (see separate entry), the Cougar and Super Puma have more powerful Turbomeca Makila turboshaft engines driving a redesigned gearbox. These gave better performance with reduced maintenance and fuel consumption.

The first production aircraft, the AS332B, used the same short fuselage as the original Puma, which gives it a 150kg (330lb) weight advantage over the later stretched aircraft. To distinguish it from its predecessor the Super Puma has a more pointed nose with space for a weather radar and a ventral fin under the vertical stabilizer. Military variants could be fitted with features such as armoured and crashworthy seating, defensive aids suites and exhaust diffusers. Abu Dhabi, Brazil and Spain were early operators of the AS332B, the latter using it in the SAR role for which it had enlarged sponsons and emergency flotation gear. Brazil, Chile and Saudi Arabia ordered the AS332F Frégate naval version of the short fuselage aircraft. This gained a folding main rotor and tail boom, under nose search radar, and provision for a dipping sonar that lowers through a cut-out in the floor where the external load hook is normally located. Armament options include lightweight torpedoes and AM.39 Exocet anti-ship missiles, which occupy almost the entire length of the fuselage forward of the undercarriage sponson when carried.

AS332M
The AS332M has a 0.76m (2ft 6in) fuselage stretch allowing an extra row of seating to be fitted, permitting 25 troops to be carried. These can be distinguished from the AS332B by the presence of six cabin windows on each side versus the five of the earlier aircraft. The performance hit from the extra weight of the longer fuselage led to slower sales for the AS332M although Brazil, Chile, Greece and Turkey have all acquired versions. France meanwhile used the AS532UL variant to carry the HORIZON battlefield surveillance radar. Folding down from the rear fuselage this I-band radar could detect moving ground targets out to 200km (124 miles). Introduced after the successful performance of a concept demonstrator during the 1991 Gulf War, four aircraft were modified and served until 2008.

H225M Caracal
Dimensions: Length: 19.5m (64ft); Rotor diameter: 16.2m (53ft 2in); Height: 4.6m (15ft 1in)
Weight: 11,200kg (24,692lb) maximum take-off
Powerplant: 2 × Turboméca Makila 2A1 turboshaft engines, 1776kW (2382shp) each
Maximum speed: 324km/h (201mph)
Range: 920km (570 miles)
Service ceiling: 6100m (20,000ft)
Crew: 2
Armament: 7.62mm (0.3in) FN MAG machine guns, 70mm (2.75in) rocket pods, MU90 torpedo and Exocet anti-ship missiles

Eurocopter AS 532UL Cougar
The Spanish Army operates 12 AS532UL Cougar, and six AS532AL. ET655 was marked as HU.21L-55 when it displayed at the International Air Tattoo in 1999. She would subsequently deploy to Afghanistan in 2005 and Iraq in 2020 as HT.27-07.

A further stretch came with the AS332L2 Super Puma MK2, or AS532U2 Cougar MK2. This added a 0.55mm (22in) plug to the aft fuselage and can be distinguished by the box-like fairing under the rear of the aircraft. Performance is restored with the use of the 1569kW (2104shp) Makila 1A2, allowing the carriage of 29 troops or 12 stretchers. While Germany and the Netherlands acquired the type as a standard utility helicopter, Saudi Arabia ordered 12 for Combat SAR. This latter variant has an under nose FLIR turret, exhaust suppressors, an extendable air-to-air refuelling probe, and provision for gun and rocket pods.

Superseded by the Caracal

At the turn of the century the AS532U2 was superseded by the EC725 Cougar MK2+, now known as the H225M Caracal. To improve performance more powerful 1776kW (2382shp) Makila 1A4 engines are mated to a reinforced gearbox, driving a new five-bladed main rotorhead. The opportunity was also taken to revise the cockpit with multifunction displays replacing all analogue gauges and installation of a four-axis autopilot system. As with the earlier Cougar models the Caracal has been developed in a range of configurations including CSAR, with similar features to the Saudi Cougars, general utility, SAR and maritime models for anti-submarine and anti-surface warfare again capable of using the Exocet missile. Several Cougar operators are replacing their earlier aircraft with the Caracal including Brazil where licence production is undertaken by Helibras for all three branches of the armed forces.

Airbus-Eurocopter AS532 Cougar helicopter of the Swiss Air Force at Alpnach, Switzerland in 2024. Switzerland operates nine Cougars which underwent a modernisation programme from 2017 to extend their operational life to 2035.

TRANSPORTS AND HEAVY LIFTERS

While helicopters are rarely the most efficient way of carrying a large payload, especially when compared to a fixed-wing aircraft, their ability to deliver it directly to its destination compensates for the extra cost. At the apex are the CH-47 Chinook, whose maximum payload has more than doubled through its production, and the Mi-26 Halo, which can deliver the same cargo as a C-130 Hercules almost anywhere on the planet.

This chapter includes the following helicopters:

- Aérospatiale SA 330 Puma
- Changhe Z-8
- AgustaWestland AW139
- Mil Mi-8
- Mil Mi-17
- Mil Mi-171
- Mil Mi-26
- Mil Mi-38
- Bell Boeing V-22 Osprey
- Boeing CH-47 Chinook

Left: US Army CH-47 Chinook helicopters line up along the flight line at Bagram Airfield, Afghanistan, June 2007.

TRANSPORTS AND HEAVY LIFTERS

Aérospatiale SA330 Puma

Developed to fulfil a French Army requirement for a transport helicopter able to carry up to 20 troops in all weathers, Sud Aviation began work on the Puma in 1963. After becoming part of Aérospatiale in 1970, they continued production, eventually developing it into the Super Puma and Cougar (see separate entry).

The resulting aircraft had a streamlined fuselage and retractable undercarriage that gained it a reputation as a fast and agile assault helicopter. Power was provided by two 884kW (1185shp) Turbomeca Turmo engines driving a four-bladed main rotor and five-bladed tail rotor. With the engines and gearbox mounted on top of the fuselage and the fuel tanks under the floor, the cabin is clear and unobstructed. Seating for troops is in four rows, two running along the side of the fuselage facing inwards with the other two placed back-to-back down the centre facing outwards. Between 14 and 20 troops can be carried depending on the configuration. Alternatively, the cabin floor has been strengthened to allow cargo to be carried internally, while a cargo hook can be fitted for underslung loads up to 2.5 tonnes in weight.

The first prototype flew in 1965 and the French Army took delivery of its first aircraft four years later. In 1967 the Puma was included in the Anglo-French helicopter agreement that saw Westland produce it and the Gazelle for the UK while Aérospatiale produced the Lynx for the French Navy. The SA330E was the designation for Pumas built for the RAF, which were essentially identical to the French Army's SA330B and the SA330C export model, although with the 984kW (1337shp) Turmo IIIC engines. The SA330D was a stillborn proposal for a Royal Navy version that lost out to the Sea King.

Eurocopter AS332L Super Puma
The People's Liberation Army Air Force (PLAAF) operate six AS332L in a non-tactical transport role in this smart blue-and-white scheme. Three EC225 are also used for VIP transport.

Upgrades

Upgrades to the basic design have been steady with enlarged undercarriage sponsons, allowing the carriage of more fuel, and polyvalent intake filters to prevent sand and dust entering the engines, which were introduced in the late 1970s. The SA330L was the first military variant with composite rotor blades that were retrofitted by many operators, including the RAF, who also acquired an additional aircraft from Argentina during the Falklands conflict.

More powerful engines have been progressively fitted; Portugal and the UK ultimately modified their aircraft to take the Makila engines used in the Super Puma. In the latter case, the resulting Puma Mk2 was able to carry twice the payload over three times the distance of the Turmo-powered Mk1.

Alongside production in France and the UK, licence-built aircraft were also made by IPTN (Indonesian Aerospace) for the Indonesian armed forces. Since 1977 Romania's Industria Aeronautica Romana (IAR) has had a licence to build Pumas, based on the SA330L standard under the IAR330 designation. As well as 89 for the country's armed forces, it has produced a similar number for export, often to countries that it would have been more difficult for Aérospatiale to export to, including South Africa, Ethiopia and Sudan.

Armament extras

IAR also developed a number of armament options with external pylons positioned outboard of the undercarriage sponsons able to carry a range of missiles, rockets and gun pods. South Africa's Atlas Aircraft reverse engineered the Puma and upgraded it to make the Oryx as a way of further evading the arms embargo against it in the 1980s. This also formed the basis of the Rooivalk attack helicopter that shares the same engines and transmission system.

SA330H Puma
Type: Utility Helicopter
Dimensions: Length: 18.15m (59ft 7in); Rotor diameter: 15m (49ft 3in); Height: 5.14m (16ft 11in)
Weight: 7000kg (15,430lb) maximum take-off
Powerplant: 2 × Turboméca Turmo IVC turboshaft engines, 1175kW (1575shp) each
Maximum speed: 273km/h (169mph)
Range: 580km (360 miles)
Service ceiling: 4800m (15,750ft)
Crew: 3
Armament: 7.62mm (0.3in) FN MAG machine guns, 20mm (0.78in) cannon

Aérospatiale SA330L Super Puma
Chile still operates two, from an original order of three, SA330L Pumas, which feature composite main rotor blades, allowing operations at higher weights than earlier aircraft. The Chilean Army also operates ten AS352 Cougars.

TRANSPORTS AND HEAVY LIFTERS

Changhe Z-8

After acquiring 12 Aérospatiale Super Frelon between 1975 and 1977, China went on to locally produce its own version, the Changhe Z-8.

Both types use three engines to drive a six-bladed main rotor and feature a large fuselage with a stern loading ramp and boat hull to allow water landings. Although Aérospatiale ceased production of the Super Frelon in 1981, Changhe have continued to develop the Z-8 family.

For anti-submarine warfare (ASW) the Z-8 is fitted with a surface search radar, dipping sonar and weapons pylons to carry lightweight torpedoes. The Z-8J is a naval transport variant equipped with a weather radar and EO turret while the Z-8S is a dedicated SAR version. Since 2011 the People's Liberation Army (PLA) has acquired the Z-8B transport, which differs from the naval version in not having sponsons for the main undercarriage. It is, however, fitted with ceramic armour and a defensive aids suite.

The Z-18 is a further development of the Z-8 based on the civilian AC313, which has a revised fuselage and makes extensive use of composite materials. The Z-18F is an ASW and ASuW variant with a 360° search radar mounted on the nose, dipping sonar, sonar buoy dispensers, and the ability to carry torpedoes and anti-ship missiles. From 2009 the Z-18J was developed for AEW duties with an AESA radar that folds down from the rear of the fuselage. The Z-18Y is the People's Liberation Army Navy's (PLAN) transport version while it has also been developed for the PLA as the Z-8G for operations at high altitude in Tibet.

Seen for the first time in 2017, the Z-8L reverts to the earlier designation

Change Z-8K
51711 is one of a number of improved Z-8K acquired from 2007 for SAR. A terrain following radar is fitted under the main radar and a FLIR turret is fitted under the right-hand side of the cockpit.

TRANSPORTS AND HEAVY LIFTERS

The Z-8L features a more modern streamlined fuselage than earlier variants of the aircraft. The large sponsons are clearly visible in this view, as well as containing fuel they are used as mounting points for chaff and flare dispensers.

Z-8K
Dimensions: Length: 23.5m (75ft 7in); Rotor diameter: 19m (62ft 4in); Height: 7m (23ft)
Weight: 13,800kg (30,360lb) maximum take-off
Powerplant: 3 × WZ-6C turboshaft engines, 1128 kW (1512 shp) each
Maximum speed: 336km/h (209mph)
Range: 1000km (600 miles)
Service ceiling: 9000m (29,500ft)
Crew: 2
Armament: 12.7mm (0.5in) machine guns

but features a significantly updated fuselage. Long sponsons on either side of the fuselage have allowed the relocation of the under-cabin fuel tanks, increasing the cabin's capacity with an internal height of 1.95m (6.4ft) and width of 2.5m (8.2ft) versus the 1.83m (6ft) and 1.9m (6.2ft) of the earlier aircraft. Despite this the new fuselage appears more streamlined due to the lower overall height. The Z-8L is in production for the PLA and a naval variant is believed to be in development for the PLAN.

Change Z-8KA
The Z-8KA performs a similar role to the Z-8K and was acquired by the PLAAF at around the same time. Seen here in an all over olive green it is also operated in the blue and white scheme seen on the opposite page.

TRANSPORTS AND HEAVY LIFTERS

AgustaWestland AW139

Launched in 1997 as a joint venture between Agusta and Bell as a replacement for the UH-1 family of helicopters, since 2005 the AW139 programme has been owned by AgustaWestland's successor Leonardo Helicopters.

The first aircraft took flight in 2001 and the commercial variant entered service in 2003, proving popular in the oil and gas industry. In 2011 the AW139M was introduced as a dedicated military variant.

The AW139 is a twin-engine aircraft with a five-bladed fully articulated main rotor, a four-bladed tail rotor and capacity for up to 15 people in the cabin. For the military market the glass cockpit is NVG compatible while a range of equipment can be fitted depending on customers' specifications. This can include search radars, rescue hoists, electro-optical turrets and ice-protection systems. The Italian Air Force was the launch customer for the AW139M in 2012 when they acquired it as the HH-139A for the SAR role. These are fitted with an under nose EO-turret, rescue hoist and infrared searchlight. A further batch of 17 HH-139B was delivered from

The National Nuclear Security Administration operates two AW139, these are equipped with radiation detection systems to provide support in the event of an incident or accident. The aircraft were delivered in 2024 replacing two Bell 412.

2020 with upgraded systems while four VH-139A are used for VIP transport.

Due to demand and in addition to the Italian production line, Leonardo established a line for the USA in 2007 while PZL-Świdnik of Poland produces bare airframes. A joint venture with JSC Russian Helicopters produced aircraft for the Russian market, however, this was suspended in 2022 following the invasion of Ukraine.

In 2018 it was announced that the MH-139 would replace the US Air Force's (USAF's) remaining UH-1N helicopters used to provide security to nuclear missile sites. Airframes are assembled by Leonardo on its Philadelphia production line before

AW139

Type: Medium Transport Helicopter
Dimensions: Length: 16.66m (54ft 8in); Rotor diameter: 13.80m (45ft 3in); Height: 4.98m (16ft 4in)
Weight: 7000kg (15,000lb) maximum take-off
Powerplant: 2 × Pratt & Whitney Canada PT6C-67C turboshaft engines, 1142kW (1531shp) each
Maximum speed: 310km/h (193mph)
Range: 1061km (659 miles)
Service ceiling: 6096m (20,000ft)
Crew: 2
Armament: Nil

USAF specific equipment is integrated by Boeing. In comparison to the aircraft it replaces, the MH-139A, also known as the 'Grey Wolf', can cruise 50 per cent faster, fly 50 per cent further and has a 30 per cent larger cabin.

Mil Mi-8

In its 60th year of series production, the Mi-8 Hip family began life as a simple transport helicopter but has since turned its hand to everything from attack helicopter to electronic warfare.

The Mi-8 was first proposed by the Mil design bureau in 1959 as a twin-turbine replacement for the piston-powered Mi-4. Although the Soviet military were not persuaded the Mi-4 needed replacing, Mil was able to argue that they were merely updating it with a turbine powerplant, which would incidentally require a redesign of the forward fuselage. The first single-engined prototype flew in 1961 powered by a 2010kW (2732shp) turboshaft; the third prototype was the first to be equipped with two purpose designed 1120kW (1522shp) Isotov TV2 turboshafts and laid the foundation for what became the world's most-produced helicopter family. Mass production commenced in 1965 at Kazan but stepped up a few years later as the Soviet military observed the US Army's massed use of the Bell UH-1 in the Vietnam War. This demand contributed to the setting up of a second production line at Ulan-Ude in 1970. Work on the improved Mi-8M started in the late 1960s with the aim of developing an aircraft with better high-altitude performance. Leveraging work done to produce the Mi-14, essentially an anti-submarine warfare derivative of the Hip, the more powerful 1400kW (1903shp) TV3-117MT engines, transmission and rotors from the newer design replaced the original equipment. Completed in 1975 the first prototype showed a significant improvement in rate of climb and service ceiling; it also gained the ability to carry three external stores pylons on each side compared to two on earlier aircraft. Entering production in 1977, the new design was designated Mi-8MT and its improved performance proved highly beneficial during the Soviet intervention in Afghanistan from 1979.

Design differences

The new aircraft appeared at the Paris Air Show in 1981 and was designated Mi-17 for the export market, although with a number of detail differences from the original Mi-8. Most obviously the tail rotor is on the left-hand side of the vertical stabilizer and rotates in the opposite direction with the front blade moving up against the downwash of the main rotor. With lift proportional to the square of the speed air flows over an aerofoil, this significantly increases the force produced by the tail rotor giving better yaw control. Additionally, the earlier Hips have circular exhausts

Mil Mi-8AMTSh-VN

Dimensions: Length: 25.3m (83ft); Rotor diameter: 21.1m (69ft 3in); Height: 5.65m (18ft 6in)
Weight: 13,500kg (29,762lb) maximum take-off
Powerplant: 2 x 1800kW (2413shp) Klimov VK-2500-03 turboshaft engines
Maximum speed: 280km/h (174mph)
Range: 580km (360 miles)
Service ceiling: 6000m (19,700ft)
Crew: 3
Armament: Up to 4000kg (8818lb) of stores including 80mm (3.3in) rocket pods, 500kg (1100lb) bombs, UPK-23-250 23mm (0.9in) gun pods, AT-9 anti-tank guided missiles

Mil Mi-8MTV-1

As well as serving the Russian armed forces, the Mi-8 is also extensively used by government agencies. RA-25226 belongs to the Chita unit of Russia's Forest Aviation Protection Service based at Chita-Kashtak airfield, Siberia.

TRANSPORTS AND HEAVY LIFTERS

compared to the more oval ones on the later aircraft and lack dust shields in front of the engine intakes. As well as the six external pylons carried on an agricultural-looking arrangement of tubes, some aircraft are equipped with a machine gun on a flexible mount in the nose.

Operations in Afghanistan demonstrated that the Hip-H, as NATO designated the new model, was a dependable workhorse in harsh mountain conditions. As well as operating as a transport it served as a gunship, convoy escort, combat search and rescue, medical evacuation, and security patrol aircraft. While the Mi-8MT proved able to continue operating with heavy battle damage, measures were taken to improve its survivability with external armour plate, and extra protection for the fuel tanks, control runs and engines. To counter surface-to-air missiles, engine infrared suppressors, chaff and flares, and an infrared jammer were also added. From 1986 the TV3-117VM engine was installed to give greater high-altitude performance.

In 1990 the Ulan-Ude plant began production of the Mi-8AMT, equivalent to the Mi-8MTV built by the Kazan factory, which is designed for hot-and-high work with a ceiling of 6000m (19,685ft). The Mi-8MTV-5 meanwhile introduced a re-profiled 'dolphin' nose that houses a weather radar and replaces the removable clamshell doors at the rear of the cabin, with a loading ramp giving the rear of the fuselage a less rounded profile than before.

The Hip has formed the basis for a variety of electronic warfare (EW) aircraft and the current version in service is the Mi-8MTPR-1 based on the Mi-8MTV-5-1 airframe. Externally the EW aircraft have the newer 'dolphin' nose, but retain the earlier clamshell rear doors to the cabin. On either side of the rear fuselage is a rectangular radome that covers an array of transmitting antenna. The window behind the main landing gear has been replaced by a chaff and flare launcher while the second window from the front of the cabin has been replaced by a conformal-receiving antenna. The cabin contains the avionics associated with the Rychag EW system with the operator's position at the front just behind the cockpit. Designed to jam enemy fire-control systems to protect friendly aircraft, once the system has detected a radar using the forward antenna the system can selectively jam up to eight radars in a single 45° sector at ranges up to around 160km (994 miles). Capable of automatic, semi-automatic or manually

The US Army operates a number of Mi-8 acquired from former Soviet air forces. This example is taking part in the Roving Sands exercise in New Mexico in 1999, providing realistic opposition forces for the US and its allies.

Mil Mi-8MT

One of six 'Hips' operated by the 301st Helicopter Transport Squadron of the Macedonian Air Force and Air Defence Forces, 306 was one of two Mi-17 (the export version of the Mi-8MT) to be upgraded by LOM Praha, in 2024.

controlled operation, the Mi-8MTPR-1 has seen service in Syria and Ukraine, six having been lost in the latter conflict by the end of 2024.

1990s upgrades

Given the extended production run it is not surprising that the age of the design has begun to become an issue. Some of the components were originally designed for the Mi-4 which first flew in 1952, while the material used for fuel and hydraulic seals has been of poor quality and led to a high maintenance burden. To rationalize its fleet and address some of these issues, Russia initiated an upgrade programme in the late 1990s to produce the Mi-8MTKO which first flew in 1999. Internally the cockpit has been made NVG compatible and upgraded through the addition of two multifunction displays to supplement the analogue gauges. A distinguishing feature is the UOMZ GOES-321VMI electro-optical (EO) turret mounted beneath the right-hand side of the cockpit; this houses an infrared camera and a laser rangefinder and target designator. The aircraft are regularly seen with armour fitted to the sides of the cockpit and have the original rounded nose.

From 2010 the Russian Air Force started taking delivery of the Mi-8AMTSh produced in the Ulan-Ude plant and powered by the 1800kW (2447shp) Klimov VK-2500 engine. A 'dolphin' nose is used, which allows a weather radar to be fitted, and most aircraft have a loading ramp rather than clamshell doors for access to the cargo bay. For the attack role, guided and unguided weapons can be carried on the external pylons including Spiral and Spiral-2 anti-tank missiles. In the transport role the Mi-8AMTSh can carry up to 37 troops or 4 tonnes of cargo. The cockpit is now equipped with four multifunction displays with analogue gauges relegated to use as standby instruments. In 2014 the Mi-8AMTSh-V debuted, which reverted to the clamshell door-style rear doors and uses new lighter metal-ceramic armour as well as upgrading the self-protection suite. The L370-8 Vitebsk-8 system features a missile approach

Mil Mi-8AMTSh-VN

Dimensions: Length: 25.3m (83ft); Rotor diameter: 21.1m (69ft 3in); Height: 5.65m (18ft 6in)
Weight: 13,500kg (29,762lb) maximum take-off
Powerplant: 2 x 1800kW (2413shp) Klimov VK-2500-03 turboshaft engines
Maximum speed: 280km/h (174mph)
Range: 580km (360 miles)
Service ceiling: 6000m (19,700ft)
Crew: 3
Armament: Up to 4000kg (8818lb) of stores including 80mm (3.3in) rocket pods, 500kg (1100lb) bombs, UPK-23-250 23mm (0.9in) gun pods, AT-9 anti-tank guided missiles

---

warning system, infrared jammers and six countermeasure dispensers. From a maintenance perspective, the aircraft has a 2000-hour time between overhauls, an increase of 500 hours over the original AMTSh.

Latest variant

The most recent Hip variant is the Mi-8AMTSh-VN, which first flew in 2018. An EO-turret is fitted under both sides of the cockpit, with a GOES-321MK on the left-hand side for

TRANSPORTS AND HEAVY LIFTERS

night flying and low-level navigation. On the right-hand side is an OPS-24N-1 targeting system with a laser guidance unit for Spiral-2 anti-tank missiles that can be fitted with armour piercing, thermobaric or fragmentation warheads. More advanced composite rotors with a new cross section and swept back tips have also allowed an increase in the maximum and cruise speeds to 280 and 260km/h (795 and 160 mph) respectively and an increase in payload. The tail rotor has adopted a four-bladed, broad-X configuration as seen on most modern attack helicopters to reduce the acoustic signature by 20%. The Mi-8AMTSh-VN is configured with the clamshell doors for the cargo bay, reportedly to improve stability when hovering in strong tail and crosswinds. Unlike most Mi-8, the Mi-8AMTSh-VN is also regularly seen with the boxy infrared suppression system fitted over the exhausts. With the improved capabilities of the latest version and continued development, it seems likely the Mi-8 will make it to a seventh decade of production.

An operator of a wide range of both Western and Soviet made equipment, Egypt has a range of Mi-8 and Mi-17 helicopters. These two have dropped off Egyptian Army soldiers, at Mubarak Military City, during Exercise Bright Star 01/02.

Mil Mi-8MTV-5

'White 95' of the Air Force and Air Defence Forces of the Republic of Belarus's 50th Separate Mixed Air Regiment. Belarus, reportedly has 36 Mi-8s of various subtypes. This example is equipped with the Vitebsk L370E8 self-protection system, including L370-5 IR jammers, L370-2 UV sensors and UV-26 countermeasures dispensers.

Mil Mi-17

The Mi-8MT was redesignated the Mi-17 for the export market when it was first shown to an international audience at the 1981 Paris Air Show.

This distinction has been retained, although confusingly equivalent sub-variants rarely have the same suffix, for example, the export Mi-8MTV-1 is the Mi-17-1V. For simplicity all Mi-8, Mi-17 and Mi-171 models are referred to as Hip by NATO.

The Kazan factory developed its own upgrades to the basic design marketed as the Mi-17-V2 and V3, which featured increased seating capacity, armour and, on the V3, additional weapons options. The Mi-17-V5 was the first 'third generation' Hip and introduced a rear-loading ramp in place of the clamshell doors and the more streamlined 'dolphin' nose able to house a weather radar. Seating was further increased from 30 troops in the V2 to 40 in the V5 and there were further improvements to the armour. China was an early customer for the type with the first eight being delivered in 2001 for use by the People's Liberation Army (PLA), these were followed by 25 Mi-17-V7 powered by VK-2500 engines giving better high-altitude performance for use in Tibet and Xinjiang in the west of the country. Some of these, potentially VIP aircraft, were delivered with the clamshell rear doors of the original Hip. Mi-17-V7 delivered to the PLA Air Force for SAR work have been fitted with an electro-optical turret, search ligh, and a terrain-following radar in a thimble-shaped radome under the nose.

Upgrades to the Mi-17 have followed a similar path to those for the Mi-8 with glass cockpits, targeting systems and defensive aids suites available and tailored to the customers' requirements.

Mil Mi-17-V5
Type: Multi-role / Utility Helicopter
Dimensions: Length: 25.3m (83ft); Rotor diameter: 21.1m (69ft 3in); Height: 5.65m (18ft 6in)
Weight: 13,000kg (28,660lb) maximum take-off
Powerplant: 2 x 1800kW (2413shp) Klimov VK-2500 turboshaft engines
Maximum speed: 250km/h (155mph)
Range: 580km (360 miles)
Service ceiling: 6000m (19,700ft)
Crew: 3
Armament: Up to 4000kg (8818lb) of stores including 80mm (3.3in) rocket pods, 500kg (1100lb) bombs, UPK-23-250 23mm (0.9in) gun pods and AT-9 anti-tank guided missiles

Mil Mi-17V-5
HA-5156 of the Tentara Nasional Indonesia Angkatan Darat (Indonesian Army), 31st Assault Squadron, based at Ahmad Yani Army Main Air Base, Java, in November 2020. Indonesia ordered twelve of these aircraft which were delivered from 2003.

TRANSPORTS AND HEAVY LIFTERS

Mil Mi-171

When production of the Mi-8MTV began at the Ulan-Ude Aviation Plant in 1991 it became the Mi-8AMT due to minor differences from the Kazan-built aircraft. For export the aircraft became the Mi-171, leading confusingly to up to three designations for what appear to be the same aircraft.

The Mi-8AMTSh became the Mi-171Sh and has been heavily marketed to South America, Asia and Africa as a budget-friendly alternative to the Mi-24 Hind. With six hardpoints and the ability to mount electro-optical turrets, radar and external armour, they can alternatively carry up to 24 troops making them more flexible than dedicated attack helicopters. The Mi-171Sh has been exported to countries such as Bangladesh, Nigeria, Peru and Algeria, generally with the original rounded nose. China meanwhile began taking deliveries of the type in 2021 with the later 'dolphin' nose. These are believed to be intended for Combat SAR operations with the People's Liberation Army Air Force and in addition to the usual protective measures have been seen with the President-S directed infrared countermeasures (DIRCM) system. This aims an infrared laser at the incoming missile and overwhelms its targeting sensor. When added a turret is fitted on the underside of the tail boom and on the outboard end of each weapon pylon carrier.

While most of the Mi-171 delivered to date have the original three-bladed tail rotor, the latest Mi-171E2 for China have been seen with the four-bladed broad-X tail rotor as seen on the Mi-8AMTSh. Although export aircraft have yet to be seen with the composite rotors of the Mi-8AMTSh-VN, it is presumably only a matter of time before this happens.

Mil Mi-171E

Operated by the Xinjang LH Regiment, LH911723 is based at Shule in the far west of China. Although the Mi-171 is well suited to the mountainous terrain in the border region, the regiment is likely to be a priority for replacement by the more powerful Z-20.

Mil Mi-171Sh

Type: Multi-role / Utility Helicopter
Dimensions: Length: 25.3m (83ft); Rotor diameter: 21.1m (69ft 3in); Height: 5.65m (18ft 6in)
Weight: 13,000kg (28,660lb) maximum take-off
Powerplant: 2 x 1800kW (2413shp) Klimov VK-2500 turboshaft engines
Maximum speed: 250km/h (155mph)
Range: 580km (360 miles)
Service ceiling: 6000m (19,700ft)
Crew: 3
Armament: Up to 4000kg (8818lb) of stores including 80mm (3.3in) rocket pods, 500kg (1100lb) bombs, UPK-23-250 23mm (0.9in) gun pods gun pods and AT-9 anti-tank guided missiles

Mil Mi-26

The largest production helicopter in the world, the Mi-26 Halo first flew in 1977 and was designed to meet a requirement for heavy lift aircraft for military and civil use.

For the military this included moving mobile ballistic missiles to remote locations in Siberia and Russia's far east. To achieve this Mil designed a helicopter with an eight-bladed main rotor powered by two 8500kW (11,556shp) Progress D-136 turboshaft engines. The resulting giant was capable of carrying an internal load of 20 tonnes, the same as a C-130 Hercules, or 90 troops. Despite its size it has a maximum speed of 295km/h (183mph), which is notably faster than the much smaller Mi-8.

To reinvigorate its fleet Russia began an upgrade programme for its Halos in the late 2010s, including a further 22 new-build aircraft. Based on the export Mi-26T2, the new Mi-26T2V has digital avionics, a new mission system, and a NPK90-2V flight control and navigation system that allows the aircraft to follow a pre-programmed route down to the final few hundred metres of an ILS (instrument landing system) approach. The cockpit has been upgraded with multifunction displays and NVG-compatible lighting for the pilots, navigator and engineer, while the loadmaster continues to be positioned in the cargo bay. Armour is provided for the cockpit while an integrated defensive aids suite provides additional protection.

Engines remain the same for all Mi-26. These were built in the late 1970s and early 1980s by Ivchenko-Progress in Ukraine, and after the 2014 invasion of Crimea all manufacturer support was cut off. Consequently, upgraded and new-build aircraft rely on rebuilt engines from the existing stock, including those taken from retired aircraft.

Mil Mi-26T2V

Type: Transport Helicopter
Dimensions: Length: 40.3m (131ft 4in); Rotor diameter: 32m (105ft); Height: 8.15m (26ft 9in)
Weight: 56,000kg (123,459lb) maximum take-off
Powerplant: 2 × ZMKB Progress D-136 turboshaft engines, 8500kW (11,400shp) each
Maximum speed: 295km/h (183mph)
Range: 500km (310 miles)
Service ceiling: 4600m (15,100ft)
Crew: 5
Armament: Window-mounted machine guns on some export models

Mil Mi-26T2

The first of four Mi-26T2s acquired by Jordan, 06819 from 26 squadron sports both a complex desert camouflage and a stylised falcon emblem on both sides of the nose. Others in the fleet wear overall gunship grey.

TRANSPORTS AND HEAVY LIFTERS

Mil Mi-38

Although work started on a replacement for the Mi-8 Hip family in the early 1980s, the dissolution of the Soviet Union delayed progress of the Mi-38 until Eurocopter helped develop it for the international market from the early 1990s. The first flight was only in late 2003 and certification completed in 2015.

The Mi-38 features a six-bladed main rotor with an X-shaped tail rotor, which give good handling combined with low noise and vibration levels. Despite having the same diameter main-rotor, the Mi-38 has a 2.6 tonne increase in maximum all-up weight compared to the Mi-8MT. While some prototypes flew with Pratt & Whitney engines, which were an option for export customers prior to the invasion of Ukraine, all production aircraft use Russian Klimov TV7 turboshafts. Like the Hip, the Mi-38 has a fixed tricycle undercarriage and access to the cargo hold via a loading ramp at the rear of the fuselage.

The Mi-38T military version uses all-Russian components including avionics to avoid the impact of sanctions. First flying in 2018, with deliveries to the Russian Air Force beginning in 2019, the 38T has provision for auxiliary fuel tanks in the cabin, foldable troop seats and a new communications system. While the cargo hold can seat 30 troops or 5 tonnes of cargo, there are also plans to use the aircraft for electronic warfare, medical evacuation and SAR. In addition to the Russian military there has been at least one export order for the Mi-38T from an undisclosed country in 2020.

Mil Mi-38T
Type: Transport Helicopter
Dimensions: Length: 20.28 (66ft 7in); Rotor diameter: 21.1m (69ft 3in); Height: 6.99m (22ft 11in)
Weight: 16,600kg (34,392lb) maximum take-off
Powerplant: 2 × Klimov TV7-117V turboshaft engines, 2100kW (2800shp) each
Maximum speed: 300km/h (190mph)
Range: 880km (550 miles)
Service ceiling: 6300m (10,200ft)
Crew: 2
Armament: Nil

Mil Mi-38T

The second prototype Mi-38, RA-38012 first flew in December 2010 powered by Pratt & Whitney Canada PW127 engines. The same month would see it fly the 800km (500 miles) from Kazan to Moscow. It was subsequently repainted in a red-and-black scheme before being withdrawn from use.

Seen here in 2007 RA-38011 was the first Mi-38 prototype, this and the second prototype used Pratt & Whitney Canada PW127 engines. From prototype 3 these were replaced with Klimov TV7-117V engines.

Bell-Boeing V-22 Osprey

The requirement for what would become the V-22 Osprey emerged after the failure of the Iran hostage rescue operation in 1980.

It was determined that what was needed for the long-range recovery of personnel was an aircraft with the speed and range of a transport aircraft, but which was able to take off and land like a helicopter. With interest from all four services, a programme was established in December 1981, although the US Army withdrew the following year due to cost concerns. Ultimately, the only bid received was from a combined Bell-Boeing Vertol team based on Bell's XV-15, which had flown in 1977.

Prototype

Full-scale development began in 1986 with the first prototype flying three years later in March 1989. Although testing initially progressed smoothly, the crash of the fifth prototype in 1992 in front of Department of Defense officials due to a fire in the right-hand engine led to an 11-month grounding. During this period there was a substantial redesign including the addition of a titanium firewall and nacelle drainage to prevent the pooling of flammable liquids from leaks.

Proprotors

The Osprey broadly resembles most medium transport aircraft with a boxy fuselage, a loading ramp to the rear of the cargo hold and sponsons for the main undercarriage. While the high-wing configuration is also typical, the major difference is the positioning of the two engine nacelles on the tips, and the use of over-sized propellers known as proprotors. Although too big to be rotated fully forwards on the ground, these are key to the V-22's unique abilities. With the nacelles rotated between 85° and 96° the aircraft is in vertical take-off or landing (VTOL) mode and operates like a helicopter.

Using a trim wheel on the throttle the pilot can rotate the nacelles, vectoring the thrust in a similar way to the Harrier, and accelerate into forward flight. Once at the fully forward or 0° position the Osprey performs like a conventional aircraft. The two engines are connected via a driveshaft through the wing that allows either engine to turn both proprotors in the event the other fails. Although unable to make a

MV-22B

Type: Tiltrotor Transport
Dimensions: Length: 17.48m (57ft 4in); Wingspan including rotors: 25.77m (84ft 7in); Height: 5m (17ft 8in)
Weight: 24,948kg (55,000lb) short take-off
Powerplant: 2 × Rolls-Royce T406-AD-400 turboshaft engines, 4590kW (6150shp) each
Maximum speed: 565km/h (316mph)
Range: 1628km (1012 miles)
Service ceiling: 7600m (25,000ft)
Crew: 4
Armament: 1 × 7.62mm (0.3in) removable M240 machine gun or 12.7mm (0.5in) M2 Browning machine gun on ramp

Bell-Boeing CMV-22B Osprey

Fleet Logistics Multi-Mission Squadron 30 (VRM-30) was the first unit to operate the US Navy's CMV-22B, gaining its first aircraft in 2020. This was followed by a deployment on the USS *Carl Vinson* to the Pacific Ocean in 2021.

TRANSPORTS AND HEAVY LIFTERS

vertical landing in this state, a running landing is possible with the proprotors designed to break away from the fuselage when they impact the ground. For shipboard stowage the V-22's proprotors fold back along the engine nacelle and the main wing then rotates 90° to lie on top of the fuselage.

Grounded after crashes

After two fatal V-22 crashes in 2000, the fleet was grounded for a full review, which led to engineering changes but found no inherent design flaws. Full production and military use were approved in September 2005 and Marine Medium Tiltrotor Squadron 263 (VMM-263) became the first operational MV-22B unit to deploy in 2007. The US Marine Corps used Ospreys extensively in Iraq and Afghanistan, where their speed and range made them the transport of choice due to their ability to shrink the operating theatre. While both the Osprey and its predecessor, the CH-46, can nominally carry 24 troops, the Osprey can cruise at 445km/h (276mph), 180km/h (112mph) faster than the CH-46's maximum speed. The Osprey also has a combat radius of 600km (372 miles) versus the 300km (186 miles) of the CH-46 essentially allowing it to move twice as much, twice as far and twice as fast as the aircraft it replaced.

USAF deployment

The US Air Force (USAF) also took delivery of their first operational CV-22B in 2007 with the 58th Special Operations Wing. Their first operational deployment saw them flying non-stop from Florida to Mali in 2008. USAF CV-22s are also based in the UK at RAF Mildenhall and in Japan at Yokota Air Base. The CV-22B features additional fuel tanks in the wings, an AN/APQ-186 terrain-following radar on the nose and

Bell-Boeing MV-22B Osprey

Marine Medium Tiltrotor Squadron 268 (VMM-268) 'Red Dragons' converted from the CH-46E Sea Knight in 2014. Home based in Kaneohe Bay, Hawaii, it is part of Marine Aircraft Group 24 (MAG 24) and provides assault support transport of troops and supplies.

directional infrared countermeasures. In 2013 after three CV-22B came under fire during an evacuation mission in South Sudan, the aircraft flew 800km (497 miles) with fuel and hydraulic leaks. In response Air Force Special Operations Command developed removable armour plates for the floor.

The US Navy has begun to acquire the CMV-22B as a replacement for the C-2 Greyhound carrier onboard delivery (COD) aircraft. While the Osprey can carry similar loads in a similar distance, it offers advantages in being able to deliver stores directly to ships other than aircraft carriers. The main adaptations for the COD role are additional fuel stored in enlarged sponsons, long-range communications equipment and a public address system for the passenger compartment.

Although in operational service, the Osprey still has a reputation for poor safety with 14 fatal accidents since it began frontline operations. A

MV-22B

Type: Tiltrotor Transport
Dimensions: Length: 17.48m (57ft 4in); Wingspan including rotors: 25.77m (84ft 7in); Height: 5m (17ft 8in)
Weight: 24,948kg (55,000lb) short take-off
Powerplant: 2 × Rolls-Royce T406-AD-400 turboshaft engines, 4590kW (6150shp) each
Maximum speed: 565km/h (316mph)
Range: 1628km (1012 miles)
Service ceiling: 7600m (25,000ft)
Crew: 4
Armament: 1 × 7.62mm (0.3in) removable M240 machine gun or 12.7mm (0.5in) M2 Browning machine gun on ramp

crash in 2023 due to a failed pinion gear led to a four-month grounding. Although revolutionary in its capability, the Osprey's ongoing troubles can no longer be excused as teething issues and have undoubtedly contributed to its limited foreign sales, with the Japanese Ground Self-Defence Force the only non-US operator.

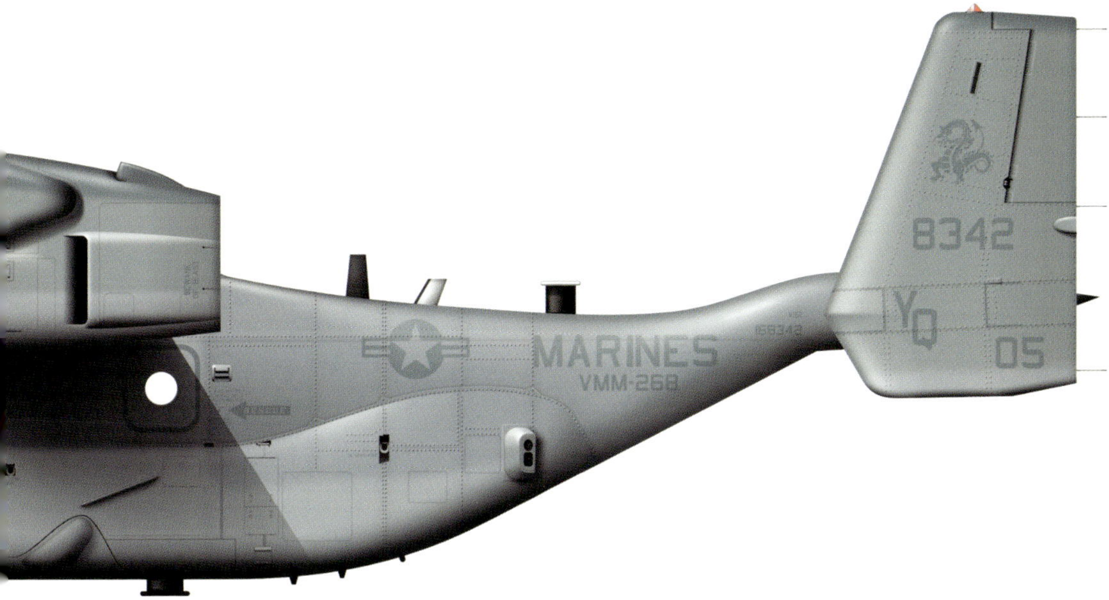

Two CV-22 Osprey of the 8th Special Operations Squadron (8th SOS) approach an MC-130H Combat Talon II of the 15th SOS to conduct air-to-air refuelling during a training mission at Hurlburt Field, Florida, 2016. Both squadrons are part of the USAF's 1st Special Operations Group, which conducts counterterrorism, combat SAR, and other special operations missions.

TRANSPORTS AND HEAVY LIFTERS

Boeing CH-47 Chinook

Work began on what became the CH-47 Chinook in 1958 after the US Army had evaluated the Vertol V-107 and found it too small.

While the V-107 would become the CH-46 and be used extensively by the US Marine Corps, Vertol worked on the larger V-114 to better meet the Army's requirement. Five were ordered under the designation YHC-1B, the 1A being the V-107, and construction was underway when Vertol became part of Boeing in March 1960.

The resulting aircraft set the mould for what has become an icon of military aviation and that has remained remarkably unchanged through its 65-year production run. Powered by two 1447kW (899mph) Lycoming T55 engines mounted either side of the rear rotor pylon, the two rotors are positioned at either end of the fuselage and geared to intermesh. This gives the Chinook the widest acceptable range for the centre of gravity of any helicopter, enabling greater flexibility when loading cargo. The 15.54m (51ft) fuselage is essentially a box for the carriage of stores and personnel with a cabin that measures 9.14 x 2.51 x 1.98m (30x 8.2 x 6.4ft). Fuel is carried in sponsons on either side of the fuselage rather than under the cabin, with the lower floor allowing easier access. A four-point undercarriage provides good stability on uneven and sloping surfaces. An indication of the efficiency of the CH-47's configuration is that its fuselage is only 6mm (0.2in) longer than that of the UH-60 Black Hawk, despite having a cabin over twice as long and roughly a third wider and taller.

Heavy lift capability

While the Chinook's design is perfectly suited to the transportation role, the use of two contra-rotating rotors also has aerodynamic advantages. All the available power can be used for lift, with the torque reactions from the two rotors cancelling each other out and removing the requirement for an anti-torque tail rotor. They also allow take-off and landing to be conducted with minimal regard for the wind direction, unlike conventional helicopters that predominately operate into wind, which restricts their approach and departure options.

Redesignated the CH-47A in 1962 as part of a tri-service alignment of designation systems, the early Chinooks had a maximum all-up mass of 15 tonnes. This allowed 4.5 tonnes of cargo to be carried either internally or externally on the single cargo hook; alternatively 33 troops or 24 stretchers could be carried. The first Chinooks arrived in Vietnam in 1965 where their heavy lift capability was used to place artillery batteries in the mountains and keep them supplied. It also found a role recovering downed aircraft, with 100 recovered in the first year of operations alone. It was not all positive, however. The heat and humidity reduced the payload by 850kg (1873lb) in the lowlands and as much as 1.3 tonnes in the mountains.

This was less of an issue for the short-lived ACH-47A gunship that replaced the cargo handling equipment with 907kg of (2000lb) armour, two forward-firing 20mm (0.78in) cannon, five cabin mounted 12.7mm (0.5in) machine guns, two 19 round 70mm (2.75in) rocket pods, and a nose-mounted 40mm (1.5in) grenade launcher. Four converted aircraft, also known as 'Guns-A-Go-Go', arrived in Vietnam in mid-1966 and were used to provide fire support for troops. Although broadly successful, the need for transport helicopters and the

TRANSPORTS AND HEAVY LIFTERS

A US Army CH-47 Chinook takes off to deliver food and water supplies to an airfield in Hawthorn, Texas, as part of disaster relief efforts in the wake of Hurricane Rita. Well suited to the role due to its range and payload, the only downside is the type's downwash, which can demolish already weakened buildings.

development of the AH-1 Cobra led to the end of the programme with the last aircraft retired in 1968.

Performance issues

To address the performance issues Boeing developed the B model. Strakes were added to the under side of the ramp and the rear rotor mast pylon trailing edge was squared off to improve directional stability. At the same time the engines were uprated to 2126kW (2890shp) and broader rotor blades with a new profile were fitted to absorb the extra power. This increased the payload at sea level to 6577kg (7.2t), a 42% increase over the A model. The first CH-47B arrived

Boeing MH-47E Chinook
Developed for special operations the MH-47E can be distinguished by the refuelling probe extending forwards from the right-hand side of the airframe. It is additionally fitted with a podded radar system on the left-hand side of the cockpit and a FLIR turret under the nose. The MH-47E is operated by the 160th Special Operations Aviation Regiment (SOAR).

89

TRANSPORTS AND HEAVY LIFTERS

in Vietnam in May 1967 and worked alongside the earlier aircraft, but it was itself soon supplanted by the C model. First flying in late 1967 with deliveries to the Army starting six months later, this model built on the changes made to the CH-47B with the engines further upgraded to 2797kW (3802shp). This allowed it to meet a requirement to carry the US Army's M198 155mm (6.1in) howitzer a range of 55km (34 miles) with the maximum all-up mass increased to 21 tonnes with a cargo capacity of 8.6 tonnes or even more over shorter distances. The C model was the last version to serve in Vietnam the C and the first Chinook to be exported to Australia, Canada, the UK, Spain and Argentina. Italy meanwhile licence built them for its Army and export customers including Iran, Libya and Morocco.

Rebuilding programme

From the mid-1970s the CH-47D programme rebuilt earlier aircraft to a new standard able to lift even more. T55-L-712 engines providing up to 3039kW (4132shp) of power for take-off helped increase the maximum all-up mass to 22.6 tonnes with a payload of 12 tonnes. Fibreglass rotor blades were introduced with an increased chord to help absorb the extra power. Cone-shaped engine intake filters that had been fitted to some CH-47C aircraft became standard while a distinctive cut-out in the rear rotor pylon provides additional cooling air for the transmission. Three under fuselage cargo hooks are fitted, the central one rated for 12 tonnes and the fore and aft rated for 9, a feature first seen on the UK's C model also known as the Chinook HC Mk1. While it is not possible to use the maximum capacity of each hook simultaneously it allows out-sizes loads, such as fuel bladders, to be delivered to multiple locations in the same sortie. Additionally, suspending a single load

Above: A CH-47D Chinook from the US Army's 3rd Battalion, 2nd Aviation Regiment, based at Camp Humphreys in South Korea. As part of the 2008 Key Resolve/Foal Eagle Exercise, the 2nd Aviation Regiments aircraft were used to drop off troops, vehicles and additional supplies at the USAF's Kunsan Air Base.

TRANSPORTS AND HEAVY LIFTERS

Above: A US Army CH-47D Chinook lands to refuel near the town of Khowst, Afghanistan, before transporting soldiers to Narizah, to conduct a cordon search in that area during Operation Mountain Sweep in August 2002. The Chinook was extensively used by NATO forces in Afghanistan due to its lifting and high-altitude capabilities.

Boeing-Vertol CH-47D Chinook
Operated by the Spanish Army's BHELTRA V Batallion ET809 was originally delivered as a CH-47C before being upgraded to CH-47D standard in 1999.

CH-47D

Dimensions: Length: 30.14m (98ft 11in); Rotor Diameter: 18.0m (60ft 0in); Height: 5.69m (18ft 8in)
Weight: 22,680kg (50,000lb) Maximum Take-Off
Powerplant: 2 × Lycoming T55-L-712 turboshaft engines, 3,039 kW (4,132 shp) each
Maximum speed: 310km/h (200mph)
Range: 558km (346 miles)
Service ceiling: 6,100m (20,000ft)
Crew: 3
Armament: 3 × 7.62 mm (.308 in) removable M240 machine guns, 1 in either forward cabin window and one on the ramp

TRANSPORTS AND HEAVY LIFTERS

from multiple hooks makes it more stable in flight allowing the Chinook to fly faster, 212km/h (132mph) versus 74–110km/h, (46–68mph) without the cargo spinning or swaying.

The CH-47D

The CH-47D entered service in 1979 with the rebuilt airframes essentially 'as new' and they were given new serial numbers. Over 400 were delivered to the US Army, the last in 2002 while most foreign operators have had a similar upgrade. Some took the opportunity to fit a glass cockpit and a weather radar in a lengthened nose. Compared to the C model, the D was found to have a 56% per cent improvement in time between unscheduled maintenance and a 45 per cent reduction in maintenance hours per flying hour. The resulting reductions in maintenance costs and increase in aircraft availability provided a compelling argument to make the initial investment to upgrade aircraft. UK Chinooks were rebuilt from 1990 and designated HC Mk2; they were the first to use a full authority digital engine control (FADEC) system.

The CH-47D model also formed the basis for the first dedicated Special Forces variant, the MH-47D. Introduced in 1985, this featured upgraded communications and navigation systems, a radar warning receiver and countermeasures dispensers. A weather radar and FLIR turret were fitted along with a fixed refuelling probe on the right-hand side of the fuselage. Twelve MH-47D were produced for the 160th Special Operations Aviation Regiment and were followed by the similarly equipped MH-47E, which entered service in 1994. The E has enlarged sponson tanks and an APQ-175 mapping and terrain-following radar in a pod on the left-hand side of the

Kawasaki CH-47J Chinook

Equivalent to the CH-47D, the CH-47J was licence built in Japan with first deliveries in 1988. JG-2927 is one of 40 operated by the Ground Self Defence Force, while the Air Self Defence Force operate 16 in a lighter camouflage. Both services also operate the CH-47JA with a nose radome, FLIR, and enlarged fuel tanks.

After the October 2005 Kashmir earthquake US helicopters stationed in neighbouring Afghanistan were deployed to provide support. This CH-47D is delivering tents to Balakot in the mountainous Khyber province.

TRANSPORTS AND HEAVY LIFTERS

forward fuselage. Additionally, the front undercarriage is repositioned forward of the front cabin window rather than aft of it as on standard aircraft.

The CH-47F

First flying in 2001, the CH-47F is a remanufactured D model designed to further improve performance. To reduce vibration levels the cockpit and forward fuselage were removed and replaced with a stronger structure that utilizes monolithic milled parts eliminating rivets and the associated maintenance burden. US Army aircraft also gained the FADEC system first used by the UK, which improved engine response and torque matching while reducing the pilot's workload and improving the handling characteristics. The F model also introduced a digital automatic flight control system (DAFCS); enabling the aircraft to follow a pre-programmed route, it can also maintain a stable hands-off hover, again reducing the pilot's workload. DAFCS has also been retrofitted to aircraft of other countries and is standard on new-build Chinooks. MH-47D and E aircraft upgraded to a similar standard are designated MH-47G and can be identified by the fixed refuelling probe, larger sponsons and repositioned forward gear.

In 2020 the CH-47F Block 2 upgrade programme commenced which has enhanced the aircraft's performance in some areas while also compensating for the slowly increasing weight of the basic airframe as more equipment has been added. More powerful T55-715 engines have allowed an increase in the maximum take-off weight to 24.5 tonnes with a payload of up to 12.5 tonnes. The lift requirement was defined against a target of lifting 10 tonnes at an altitude of 1200m (4000ft) at 35°C, ensuring that heavy payloads can be moved even in adverse conditions.

To better use the available power, the flight-control system has been redesigned to provide tactile feedback of the torque split between the fore and aft rotor systems. This allows the safety margin to be safely reduced to extract more of the available performance from the aircraft.

The first of up to 465 Block 2 F models was delivered to the US Army in mid-2024. Plans for further engine upgrades have been looked at and although a replacement aircraft is included in the Future Vertical Lift programme, the Chinook is the lowest priority of the types to be replaced. Given its continued utility it seems likely the CH-47 will still be in service in 2061, 100 years after its first flight.

CH-47J

Dimensions: Length: 30.14m (98ft 11in); Rotor Diameter: 18.0m (60ft 0in); Height: 5.69m (18ft 8in)
Weight: 24,494kg (54,000lb) Maximum Take-Off
Powerplant: 2 × Lycoming T55-GA-714A turboshaft engines, 3,529 kW (4,733 shp) each
Maximum speed: 310km/h (200 mph)
Range: 740km (460 miles)
Service ceiling: 6,100m (20,000ft)
Crew: 3
Armament: 3 × 7.62 mm (.308 in) removable M240 machine guns, 1 in either forward cabin window and one on the ramp

NAVAL HELICOPTERS

Modern maritime helicopters carry extensive sensor suites, including radars and sonars, that can match the capabilities of their parent ships. The Royal Navy's Merlin anti-submarine warfare (ASW) aircraft have even been described as flying frigates, and are able to direct other aircraft and ships in pursuit of a target. Often the only aviation asset available, most of the types listed here can also conduct basic transport and search and rescue (SAR) missions with a change of role equipment to maximize their utility.

This chapter includes the following helicopters:

- Kamov Ka-27
- Kamov Ka-31
- CH-53 Sea Stallion
- AgustaWestland AW159 Wildcat
- Westland Lynx
- Sikorsky SH-60 Seahawk
- Sikorsky S-61 Sea King
- AgustaWestland AW101

An SH-60F Seahawk helicopter of Helicopter Antisubmarine Squadron Eleven (HS-11), 'Dragonslayers', landing on the aircraft carrier USS *Enterprise* (CVN 65). HS-11 has now transitioned to the MH-60S Knight Hawk and been redesignated HSC-11.

NAVAL HELICOPTERS

Kamov Ka-27

A shipborne ASW helicopter design work on the Ka-27 began in 1969 with the first flight happening four years later. Given the NATO reporting name 'Helix', the resulting aircraft has Kamov's trademark coaxial rotor system driven by two Isotov TV3 turboshaft engines.

This allows for a more compact design without the need for a tail boom, an advantage when embarking on ships as it reduces the deck space needed to operate and store the aircraft. The lack of tail rotor also makes a safer environment for the deck crew. The co-axial rotors do, however, require a higher ceiling for the hangar, which has led to some novel solutions, such as on the Udaloy class where the hangar roof slides forward so the aircraft can enter. Once in the hangar the floor moves down and forwards allowing the roof to clear the rotors as it closes.

Unusual service issue

Entering service in 1981 the Ka-27PL had a 360° Osminog search radar under the nose to aid in navigation and detecting shipping and surfaced submarines. A VGS-3 dipping sonar housed in the rear of the aircraft is lowered from a hatch just forward of the fuselage/tail junction. Designed to be operated down to a depth of 175m (246ft) and detect submarines at depths of up to 500m (1640ft) it has a detection range of around 7km (4.3 miles). Initial service use in the Mediterranean discovered an unusual issue where the operator would hear radio stations rather than the expected submarines. Fault finding discovered that unshielded wires in the cable connecting the sonar to the aircraft were acting as an antenna. For the anti-submarine warfare (ASW) role the Helix can also carry 36 sonobuoys or a homing torpedo in the under-fuselage weapons bay. Given this limitation when operating at range from the mothership, two aircraft will operate together with one in the pure search role with sonobuoys while a second will be in the search-strike role with a torpedo. Most Ka-27PL are fitted with a magnetic anomaly detector (MAD) for final location of a submerged submarine, although it is not used by

Kamov Ka-27PS Helix

'33 Yellow' wears a blue-and-white colour scheme commonly seen on Ka-27 assigned to SAR duties. Just visible in the side view is an orange stripe that runs down the centre of the engine and transmission housing and along the underside.

Ka-27

Type: ASW Helicopter
Dimensions: Length: 11.3m (37ft 1in); Rotor Diameter: 15.8m (51ft 10in); Height: 5.5m (18ft 1in)
Weight: 12,000kg (26,455lb) maximum take-off
Powerplant: 2 × Isotov TV3-117V turboshaft engines, 1660kW (2230shp) each
Maximum speed: 270km/h (170mph)
Range: 980km (610 miles)
Service ceiling: 5000m (16,000ft)
Crew: 3-5
Armament: 1 × APR-3E homing torpedo or 4 × S3V depth charges

NAVAL HELICOPTERS

A Kamov Ka-27PS from the Russian navy's *Admiral Vinogradov* operating in the Gulf of Aden in 2009. The *Admiral Vinogradov* is a Udaloy class destroyer and typically deploys with two Ka-27 which are housed in separate hangars.

aircraft of the Northern Fleet due to the proximity of the magnetic pole to their operating area.

The Ka-27PS was developed alongside the PL as a SAR model. This lost the sonar equipment but retained the radar, which along with an A-817 transponder receiver could be used to search for survivors. Given the potential requirement to conduct rescues from nuclear submarines, an X-ray meter was also installed in the cockpit. An electric hoist is fitted by the cabin door and has a capacity of 300kg (661lb). Alternatively inflatable lifebelts, boats and marker buoys can be deployed, the latter spreading a bright dye into the water to assist other aircraft and ships in locating the survivors. The cabin can be configured with medical equipment as required including four stretchers and an oxygen machine.

For the assault role the Ka-29TB can carry 16 troops with provision for rockets, guns or missiles to be carried on four hardpoints on stub wings either side of the central fuselage. A 12.7mm (0.5in) cannon is fitted behind a hatch in the nose and a 30mm (1.18in) cannon can be fitted to the right-hand side of the aircraft with ammunition fed from the cabin. Lacking the search radar the nose has a more streamlined appearance, although it is fitted with a low-light camera, infrared camera and millimetric radar for weapons guidance. Intended to operate from assault ships and aircraft carriers, the Ka-29 has also been seen operating from ashore in the Ukraine conflict.

Deployment of the Ka-28

While there was little development or deployment of the Ka-27 in the immediate aftermath of the dissolution of the Soviet Union, sales of the export model, the Ka-28, were made to several countries including India, Vietnam and China. The latter

NAVAL HELICOPTERS

received its first aircraft at the turn of the century with a follow-on order in 2009 despite the indigenous development of the Z-9C for the ASW role. This indicated the Helix has some advantages over the smaller aircraft, potentially in the areas of endurance and weapons capacity. Kamov has also had a reasonable degree of success with the Ka-32 civil version of the aircraft, which has proved popular for firefighting and as a flying crane.

Deliveries of the Ka-27M to the Russian Navy began in 2016. These are modernized Ka-27PL that are stripped down and rebuilt in an 'as new' condition. The flight and navigation system has been replaced with a new digital system based on an open architecture to ease future upgrades. The tactical system has also been digitized and is now fed by a Kopyo-A X-band radar with a detection range of 250km (155 miles) with similar improvements to the Ros-VM dipping sonar and MMS-27 MAD. Forty-six aircraft have been upgraded for the Russian Navy while a similar rebuild and modernization has been carried out on at least 10 Ka-29TB.

Kamov Ka-28
7520 from the Vietnam People's Navy, seen aboard VPN *Quang Trung* (016), a Gepard 3.9 Class frigate, in Singapore in May 2019. Vietnam ordered eight Ka-28 aircraft, the export version of the Ka-27PL, for the anti-submarine warfare role.

Kamov Ka-32T
RDPL-34077 of the Lao People's Liberation Army Air Force, based at Wattay International Airport, Vientiane, Laos in May 2012. Laos received six Ka-32Ts from 1997, one of which subsequently crashed in a fatal accident. The Ka-32T (NATO 'Helix-C') is a transport version of the Ka-27, accommodating two crew and 16 passengers and is used by both civil and military operators.

Kamov Ka-31

The Ka-31 was designed to fulfil a Soviet requirement for an airborne early warning (AEW) aircraft for its Kiev- and Kuznetsov-class carriers.

Although only two had been delivered at the time of the Soviet Union's dissolution in 1991, trials work and limited deployments still took place under Russian auspices. Development was mainly funded by orders from the Indian and Chinese navies. The former ordered its first aircraft in 1999 and by 2012 had received 14 in total. China meanwhile ordered nine in 2008 all of which remain in service.

Modified Ka-29
Based on the Ka-29 transport version of the Helix, which has a wider cockpit section, the Ka-31 loses the under-nose sensors and internal cannon. For the AEW role it gained an E-801M radar that has its antenna mounted under the fuselage, folding down to rotate at 6rpm once the aircraft is airborne and the undercarriage has retracted. Due to the size of the antenna, a SAU-37D automatic flight control system is fitted to minimize disturbance to the flight path as it rotates. The radar can detect air targets at 111km (70 miles) and surface contacts out to 250km (155 miles), which are transmitted to the aircraft's parent ship via datalink for subsequent direction of friendly forces. Unlike Western AEW aircraft, the Ka-31 has no command-and-control capability of its own.

Russia finally ordered its own production aircraft in 2008, at least one of which subsequently deployed to the Mediterranean with the Admiral Kuznetsov in 2016. Although the Russian carrier's ongoing problems have prevented further deployments, Ka-31s have also been ordered to support the Russian Army in providing overland surveillance. For this role a KRET Vitebsk self-protection suite has been added. At least one of these aircraft, reportedly designated Ka-35, has been seen operating in Syria.

Kamov Ka-31
9284 belongs to the Chinese Peoples' Liberation Army Navy Air Force (PLANAF), assigned to 4th Division of the 11th Air Regiment, Eastern Theatre Command based at Ningbo/Zhuangqaio, on the coast of the East China Sea. It is shown here with its E-801 Oko L-band radar antenna deployed beneath the fuselage and its undercarriage in the retracted position.

Ka-31
Type: AEW Helicopter
Dimensions: Length: 11.3m (37ft 1in); Rotor Diameter: 15.8m (51ft 10in); Height: 5.5m (18ft 1in)
Weight: 12,500kg (27,558lb) maximum take-off
Powerplant: 2 × Isotov TV3-117V turboshaft engines, 1864kW (2500shp) each
Maximum speed: 250km/h (160mph)
Range: 600km (370 miles)
Service ceiling: 5000m (16,000ft)
Crew: 2

NAVAL HELICOPTERS

Sikorsky CH-53 Sea Stallion

Due to concerns about the suitability of the XC-142A VTOL transport for shipborne operations, in 1962 the Bureau of Naval Weapons issued a tender on behalf of the US Marine Corps (USMC) for a maritime helicopter for ship-to-shore transportation.

Although Boeing proposed a Chinook derivative and Kamen a military version of the Fairey Rotodyne, Sikorsky were successful with the S-65. This took transmission and rotor system elements from the CH-37 and CH-54, and married them to a fuselage developed from the HH-3s. The resulting aircraft, designated CH-53A by the US military, could carry 3.6 tonnes or 38 troops internally or a six-tonne external load over a 185km (115m) radius of action. A top speed of 315km/h (196mph) made it the fastest production helicopter in the western world in 1965. The first five CH-53A were delivered to the US Marine Corps (USMC) in September 1966 with the similar HH-53B delivered to the US Air Force for Combat SAR just over a year later.

CH-53D
From 1969 the USMC started taking delivery of the CH-53D. This increased the internal cargo capacity to 6.3 tonnes thanks to an uprated transmission system and 2926kW (3925shp) T64-GE-412 engines.

Sikorsky CH-53K King Stallion
Part of the second lot of low-rate initial production (LRIP) aircraft, 17003 was built in 2022 and delivered to the USMC in December of that year. Marked as CJ 013, it is operated by Marine Heavy Helicopter Squadron 461 (HMH-461, 'Ironhorse'), based at Marine Corps Air Station New River, North Carolina.

Sikorsky CH-53E Super Stallion
Still in service in 2025, CH-53E 165504, marked as MT 04, belongs to Marine Heavy Helicopter Squadron 772 (HMH-772, 'Hustlers'), of Marine Aircraft Group 49, 4th Marine Aircraft Wing, based at Joint Base McGuire-Dix-Lakehurst, New Jersey.

CH-53K
Type: Heavy Lift Helicopter
Dimensions: Length: 30.17m (99ft 0.5in); Rotor diameter: 24.8m (79ft); Height: 8.6m (28ft 3in)
Weight: 39,900kg (88,000lb) maximum weight with external load
Powerplant: 3 x General Electric T408 (GE38-1B) turboshaft engines 5600kW (7500shp) each
Maximum speed: 315km/h (195mph)
Range: 850km (530 miles)
Service ceiling: 4900m (16,000ft)
Crew: 4
Armament: 3 x GAU-21 12.7mm (0.5in) machine guns

NAVAL HELICOPTERS

They also gained engine air particle separators (EAPS) to prevent dust and debris entering the intakes. From 1970 Sikorsky developed the improved rotor blade (IRB) to replace the original aluminium units. These used Nomex honeycomb profiles bonded to a titanium spar and wrapped in fibreglass, which allowed an increase in max all-up mass of 2.4 tonnes. These were followed by a new rotorhead in 1972 that replaced the original mechanical bearings with elastomeric ones. Constructed from a sandwich of rubber and metal, these allow the same degree of movement as the original bearings without requiring lubrication or other

This CH-53E is embarking Marines for a fast rope training exercise at Yuma, Arizona in 2015. During this manoeuvre the aircraft will enter a hover, allowing the Marines to exit by sliding down a thick rope using their hands to slow their descent.

maintenance. CH-53Ds remained in USMC service for 43 years and were finally retired in 2012.

CH-53E Super Stallion
The CH-53E Super Stallion was the result of a 1967 USMC requirement for an aircraft that could carry 1.8 times the load of the CH-53D. An evolution of the CH-53D, the fuselage was stretched 1.88m (6.2ft) and allowed internal seating for 55 troops, or 13.6 tonnes of cargo. A third engine was added behind the rotor shaft, with the intake between the transmission housing and the left-hand engine. The three engines are connected to a strengthened transmission system driving a new seven-bladed main rotor head. Rotor diameter was increased from 22.02m (72.2ft) to 24.08m (79ft) using an insert at the blade root while continuing to use the CH-53D's IRB. The tail rotor pylon was canted over by 20° to provide an element of lift that permitted a wider allowable range for the centre of gravity. To increase the range an extendable air-to-air refuelling probe was added along with provision for external fuel tanks. The maximum external load of the CH-53E is 16.3 tonnes. Entering service in 1981 the Super Stallion is fast with a top speed of over 300km/h (186mph), and manoeuvrable, despite its size.

Alongside the CH-53E, the US Navy ordered the MH-53E Sea Dragon airborne mine countermeasures (AMCM) aircraft, which entered service in 1987. The most obvious external difference is the fuselage sponsons, which are much larger on the Sea Dragon, reaching almost as high as the engine nacelles while having a more teardrop-shaped cross section. These allow for a greater fuel capacity while an additional seven ferry tanks can be fitted in the cargo hold. For the AMCM mission the aircraft can tow a range of systems through the water to trigger acoustic or magnetic mines at no risk to itself, or a side-scanning sonar

Seen in 2017 at Palm Beach, Florida, this CH-53K King Stallion was engaged in the test programme. Shortly after this flight in April 2017 Lockheed Martin announced that the CH-53K King Stallion had passed its Defence Acquisition Board assessment and was approved for low-rate initial production (LRIP).

for mine location. Many of the towed systems have their own generators to provide power, part of the reason for the additional fuel capacity in the Sea Dragon's sponsons.

King Stallion

Also known as the King Stallion, the CH-53K originated in a 2006 Sikorsky proposal as an alternative to upgrading the 53E fleet. Although originally intended to achieve initial operating capability (IOC), in 2015 this was delayed until 2022. The maximum all-up mass of the CH-53K is 39.9 tonnes with an external load compared to the E model's 33 tonnes. The fuselage has been widened by 30cm (11.8in), which allows the 463L pallets used by the C-130 Hercules and other cargo aircraft to be loaded into the cargo hold. Alternatively a Humvee can also be carried internally. External loads can be carried on one of three hooks, the fore and aft ones being able to carry 11.3 tonnes while the central one is rated at 16.3 tonnes. To maintain the same overall footprint for shipboard operations the sponsons have been reduced in width, although an increase in height and length allows sufficient internal fuel to be carried that external tanks are no longer required. The three T408 engines each deliver 5600kW (7613shp) giving 70 per cent more power. These drive new composite rotor blades that are controlled by a fly-by-wire system. The King Stallion significantly increases the lift capacity of the USMC heavy transport squadrons. The Super Stallion meanwhile is likely to remain in service until around 2030 with reserve units being the last to operate the type.

A CH-53E Helicopter attached to Marine Medium Helicopter Squadron 165 (HMM-165) onboard USS *Peleliu* off San Diego in May 2010, prior to deploying to the Persian Gulf. During this deployment the aircraft would take part in disaster relief operations in Pakistan after the monsoon led to extensive flooding.

NAVAL HELICOPTERS

AgustaWestland AW159 Wildcat

Ordered for the UK military in 2006, the Wildcat is recognizably descended from the Westland Lynx and carries over some components, most notably the main rotor gearbox, rotorhead and main rotor blades.

These are paired to a new tail rotor system while the 664kW (890shp) Rolls-Royce Gem engines have been replaced by the 1015kW (1361shp) LHTEC CTS800, giving a significant increase in power. This has increased the maximum take-off weight to 6 tonnes with the design allowing for further growth over the aircraft's life without structural alterations. As on the Lynx, the rotor can generate negative lift to assist in holding the aircraft on to a moving flight deck.

Wildcat modifications

Monolithic machined aluminium panels are used in the airframe's construction, reducing the component count by up to 80 per cent on those parts. To provide more space for avionics systems, the nose and rear fuselage have been enlarged with improved access to the latter via a hatch on the left-hand side. The new horizontal stabilizer improves the flying characteristics while incorporating conformal antenna in the vertical fins. Survivability has been significantly improved with stroking crashworthy seats for the crew and passengers. Protection is also provided by the optional fitment of armour plates and a wire strike protection system. The engine intakes include a particle separation system to prevent dust and debris affecting performance, while the exhausts are designed to mix the hot gases with rotor downwash to reduce the infrared signature. An AN/AAR-57

Wildcat HMA Mk2
Type: Anti-Surface Warfare Helicopter
Dimensions: Length: 15.24m (50ft); Rotor diameter: 12.8m (42ft); Height: 3.73m (12ft 3in)
Weight: 6000kg (13,228lb) maximum take-off
Powerplant: 2 × LHTEC CTS800-4N turboshaft, 1015kW (1361shp) each
Maximum speed: 311km/h (193mph)
Range: 463km (287 miles)
Service ceiling: 3657m (12,000ft)
Crew: 2
Armament: 1 × 12.7mm (0.5in) M3M heavy machine gun, Martlet and Sea Venom air-to-surface missiles, Sting Ray torpedoes and Mk11 depth charge

A Royal Navy Wildcat HMA Mk2 taking part in multinational operations in the Persian Gulf in July 2021. Mounted outside the open cabin door is an M3M heavy machine gun, this can be used for firing warning shots and disabling small craft.

NAVAL HELICOPTERS

The USS *Gerald R Ford*'s flight deck is significantly larger than the ones the Wildcat typically operates from. However it is not unusual to be used to transfer personnel to larger ships for planning conferences or to collect stores.

missile warning system and Vicon 78 countermeasures dispensers are fitted as part of an integrated defensive aids suite (DAS). A Wescam MX-15Di electro-optical turret above the nose contains infrared and daylight cameras and a laser designator. Naval Wildcats are also fitted with a Seaspray 7400E active electronically scanned array (AESA) radar under the nose. This provides a significant increase in detection range compared to the 1970s system fitted to the Lynx. It also provides far better target identification capabilities while having a lower probability of detection by opposing forces.

The first production aircraft were delivered to the British Army in 2012 and to the Royal Navy in 2013. These were used for operational evaluation by the respective services before formally entering service with the Army in 2014 and the Royal Navy the following year with a deployment on HMS Lancaster to the South Atlantic. The Wildcat had completely taken over from the Royal Navy's Lynx by 2017 and the Army retired their final examples the following year. Initially Navy and Army aircraft were identical apart from the fitment of the radar and deck-landing equipment and could theoretically be converted from one role to the other. However, this capability was never used and subsequent changes led to two distinct marks of aircraft.

AW159 Wildcat
It is believed that the Royal Navy operates 28 Wildcats as part of its inventory.

NAVAL HELICOPTERS

In 2018 two Wildcats of the Navy's Maritime Interdiction (MI) Flight supported Special Forces in taking the MV Grande Tema that had been hijacked by stowaways in the Thames Estuary. Wildcats were also used the following year to intercept the Iranian oil tanker Grace 1 off Gibraltar. Since 2019 an aircraft has been deployed to the Middle East with the Royal Navy's forward deployed frigate while most frigates and destroyers deploy outside UK waters with one embarked.

Armament
On entry to service the Wildcat HMA Mk2 could only be armed with the 12.7mm (0.5in) M3M heavy machine gun, Sting Ray torpedoes and Mk11 depth charge. In 2021 an air-to-surface missile capability that had been lost with the retirement of the Lynx/Sea Skua combination was reintroduced. The Thales-developed Martlet lightweight multi-role missile (LMM) is a supersonic 13kg (28lb) laser-guided weapon with a range of 8km (5 miles). Although primarily intended to counter small surface targets such as fast inshore attack craft, it has also demonstrated an anti-UAV capability. The missiles are carried in sealed tubes that are loaded into five round launchers; two launchers can be carried on either side on specialized weapons wings. More refined than the carriers used on export aircraft, the wings generate up to 360kg (793lb) of lift each, offsetting the drag generated by the launchers to achieve the required range and endurance. To counter larger targets, up to corvette size, the HMA Mk2 will also be armed with the MBDA Sea Venom with a range of 20km (12

Easily distinguished from the naval variant by the lack of radar, the Wildcat AH Mk1 is a regular visitor to Norway with 847 Naval Air Squadron of the Royal Navy's Commando Helicopter Force (CHF), who pool their aircraft with the Army.

miles) and a 30kg (66lb) warhead. In contrast to Martlet this has an inertial guidance system with an infrared seeker for terminal guidance and the option of control by the launching aircraft via a datalink to confirm the target selection and point of impact. Up to four Sea Venom can be carried or a mix of ten Martlet and two of the larger missiles. Export users have opted for the Israeli-developed Spike NLOS missile whose capabilities lie between those of Martlet and Sea Venom. They have also chosen the compact-FLASH dipping sonar, giving them an anti-submarine warfare (ASW) capability.

Westland Lynx

Development of the Lynx began in the 1960s with studies for a replacement for the Westland Scout and Wasp.

Along with the Puma and Gazelle, the Lynx was part of the Anglo-French Helicopter Agreement, although unlike the former two aircraft the UK led the programme. Development formally began in 1968 for service entry in 1975, but this slipped by a year. Power was provided by two Bristol Siddeley (later Rolls-Royce) 671kW (900shp) Gem engines, although the Pratt & Whitney PT-6 was also trialled on one aircraft. To fit within the relatively low shipboard hangars of the time while maintaining a usable cabin height, conformal gears were used in the main rotor gearbox giving it a shallow profile. Even so headroom in the back of the aircraft is limited. The gearbox drives a rigid rotorhead made of titanium with all blade movement being accommodated by flexing of the head rather than via hinges. Consequently, the Lynx is highly responsive to the controls aiding operations from small ships in high sea states. To hold the aircraft on the ship's deck, a harpoon extends from below the fuselage to lock into a stainless-steel grid. The toed-out mainwheels allow the Lynx to turn on the spot while attached to the ship and depart into wind if necessary. For onboard stowage, the main rotors and vertical stabilizer can be manually folded to reduce the length to 10.62m (34ft 10in).

An Anglo-Dutch trials unit, 700L NAS, was formed in 1976 with Lynx HAS Mk2 and conducted the first embarkation, onboard HMS Sirius, the following year beginning 40 years of embarked service in the Royal Navy. Brazil and Argentina had taken delivery of naval Lynxes by the end of the decade while France, Denmark, Germany and Norway had received aircraft at the beginning of the 1980s.

Naval variant

Initial plans for the naval variant had included sonobouy dispensers or a dipping sonar and although Royal Navy models ultimately never carried either system, many export customers aircraft did. Instead, for the anti-submarine role, torpedoes were dropped at the direction of a ship or anti-submarine warfare (ASW) helicopter. For anti-

Lynx HMA Mk8
Type: Anti-Surface Warfare Helicopter
Dimensions: Length: 15.24m (50ft); Rotor diameter: 12.8m (42ft); Height: 3.67m (12ft 1in)
Weight: 5329kg (11,750lb) maximum take-off
Powerplant: 2 × Rolls-Royce Gem turboshaft, 686kW (920shp) each
Maximum speed: 324km/h (201mph)
Range: 528km (328 miles)
Service ceiling: 3657m (12,000ft)
Crew: 2
Armament: 1 x 12.7mm (0.5in) M3M heavy machine gun, Sea Skua anti-ship missiles, Sting Ray torpedoes and Mk11 depth charge

Westland Lynx AH Mk9
ZF538 was built as an AH Mk9 and served with the Royal Marines and later the Army Air Corps. It was subsequently converted to Mk9A standard with the same LHTEC engines used on the Wildcat for improved performance; as such it would be the last Lynx to fly in UK service on the 26th of January 2018.

NAVAL HELICOPTERS

Westland Lynx HAS Mk3

Delivered as a HAS Mk2 in 1977, XZ235 would be converted to Mk3 standard in 1986. Marked as 304, it served with 815 Naval Air Squadron (NAS). In 2004, it was serving with 702 NAS as 633, where the author flew it for his first flight on the Lynx.

Cockpit
The crew of two sit side-by-side in the cockpit. The Mk3 could be configured with dual controls for pilot training. However, normally the left-hand controls were removed and the instrument panel, replaced with the radar display and controls.

surface warfare, up to four Sea Skua, sea-skimming, anti-ship missiles could be carried, guided by a Sea Spray radar fitted in the nose. The first operational use of these was by the Royal Navy during the 1982 Falklands conflict where eight missiles were launched and seven hits scored. By contrast, Argentina's two Lynxes saw little action; one was lost in an accident and the other was grounded and sold to Denmark after the end of hostilities.

Shortly after the Falklands conflict the HAS Mk3 entered Royal Navy service with an uprated gearbox and more powerful Gem Mk 204 engines. The HAS Mk3 also gained the Orange Crop ESM system to detect radar transmissions, a four-bag flotation bag system to replace the original two bag system, and a towed magnetic anomaly detector (MAD). Mk2 aircraft were eventually upgraded to the same standard and other operators used

the engine and gearbox upgrades, although the avionics fit varied by country. For Operation Granby – the UK's contribution to the liberation of Kuwait in 1991 – aircraft were modified to Mk3GM standard with provision for the Yellow Veil jamming pod, Sand Piper FLIR and IR jammers. There the Lynx repeated the success of the Falklands conflict, scoring 17 hits with Sea Skua and sinking 12 Iraqi vessels.

NAVAL HELICOPTERS

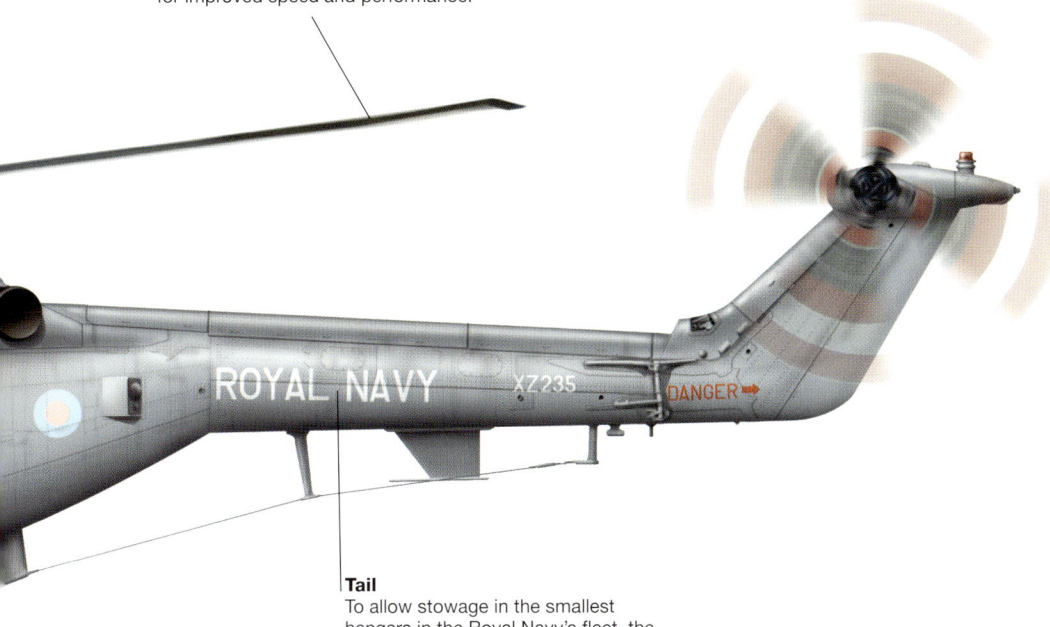

Rotor Blades
The Lynx entered service with metal blades of constant chord, from 1995 these were replaced with composite blades based on the British Experimental Rotor Programme design for improved speed and performance.

Tail
To allow stowage in the smallest hangars in the Royal Navy's fleet, the Lynx's tail boom folds just forward of the vertical stabiliser. Manually operated, a built-in ratchet handle is used to ensure the two parts are locked together prior to flight.

Maritime attack role

When the Lynx HMA Mk8 entered service in 1995, the new designation emphasized the maritime attack role over the anti-submarine role. Development had been underway since the mid-1980s with a Racal central tactical system (CTS) trialled in modified HAS Mk3s. Externally the nose was reconfigured with a 360° radome, although due to cost savings the radar still only scanned in the forward sector. Above this was a Sea Owl infrared camera that allowed target identification at increased range by day and night. Tail rotor control was improved by reversing the direction of rotation so that the upgoing blade was in the main rotor downwash, therefore increasing the force generated. At the same time composite main rotor blades (CMRB) replaced the original metal units. These were based on work done as part of the British Experimental

Lynx HAS Mk3

Type: Anti-Surface Warfare Helicopter
Dimensions: Length: 15.24m (50ft); Rotor diameter: 12.8m (42ft); Height: 3.67m (12ft 1in)
Weight: 5329kg (11,750lb) maximum take-off
Powerplant: 2 × Rolls-Royce Gem turboshaft, 686kW (920shp) each
Maximum speed: 324km/h (201mph)
Range: 528km (328 miles)
Service ceiling: 3657m (12,000ft)
Crew: 2
Armament: 1 x 12.7mm (0.5in) M3M heavy machine gun, Sea Skua anti-ship missiles, Sting Ray torpedoes and Mk11 depth charge

NAVAL HELICOPTERS

Westland Lynx SH-14D

In Dutch service the Lynx could be fitted with a dipping sonar, an option not pursued by the UK. This went in the cabin along with a third crew member, the sonar being deployed through a hole in the bottom of the fuselage.

Rotor Programme (BERP) and featured enlarged paddle tips that delayed the onset of the critical Mach number. Later upgrades included a digital signal processor (DSP) incorporated in the radar to give a track-while-scan capability, digital zoom and overlay of the tactical plot on the radar picture.

The HMA Mk8 formed the basis for the Super Lynx, which continued the type's export success with sales to Brazil, Portugal, South Korea and South Africa through the 1990s, and Malaysia in 2003. Many of these aircraft are powered by the LHTEC CTS 800 turboshaft subsequently used on the Wildcat, which along with upgraded avionics provide the older airframe with similar capabilities. Still in service with nine countries in 2025 and with Portugal receiving five upgraded Mk95A as recently as 2021, the Lynx is likely to remain operational well into the 2030s.

Originally delivered to the RN in 1980 as a Lynx HAS Mk2 XZ723 was subsequently upgraded to a HAS Mk3 and then in the 1990s to HMA Mk8 standard. Seen here at the International Air Tattoo in 2014 it was finally retired in 2016.

NAVAL HELICOPTERS

Sikorsky SH-60 Seahawk

In the early 1970s when the US Navy (USN) was looking to develop a replacement for its Sea Sprite helicopter, it chose the Black Hawk for its size, development potential and cost-saving opportunities. The SH-60 first flew in December 1979, shortly after the Black Hawk itself entered service.

Marinization led to a number of changes along with those required to perform the SH-60B's primary and secondary missions. Most obviously the tail wheel was moved forwards to the fuselage/tail junction, reducing the footprint and improving the margin for error when operating to the small flight deck of a frigate or destroyer. An electric fold system was developed for the main rotors while the horizontal stabilizer folds upwards, allowing the tail pylon to be folded forwards and reducing the space required to store the aircraft onboard ship.

Engine power was increased from 1150kW (1542shp) to 1260kW (1690shp), which was accompanied by strengthening of the transmission system with larger gear faces and thicker driveshaft walls. These improvements ultimately found their way to the UH-60L version of the Black Hawk. To operate from a ship in poor weather, the Seahawk uses the Recovery, Assist, Secure and Traversing (RAST) system. In use the aircraft deploys a cable down to the ship's deck to in turn recover the haul down cable that is attached to the underside of the fuselage. Once this is secured to the aircraft, a shipboard winch pulls the aircraft down onto the deck with a force of around 2 tonnes. Once on deck the traversing system can move the helicopter into the hangar after the automatic blade fold is completed.

For the anti-submarine warfare (ASW) mission a 25-round sonobuoy dispenser was installed on the left-hand side of the cabin, replacing the cabin door and a towed magnetic anomaly detector (MAD) was housed on a pylon on the right-hand side of the fuselage. For anti-surface warfare the SH-60B was equipped with a 360° search radar under the forward fuselage and electronic support measures (ESM) antenna to detect radar transmissions. These allowed the Seahawk to locate shipping and pass back targeting information to its parent ship. Information from all the sensors was processed by a central computer developed by IBM and could be transmitted via datalink.

Sikorsky SH-60B Seahawk

Helicopter Antisubmarine Squadron Light 42 (HSL-42) operated the SH-60B from 1984 until 2013, when it was redesignated Helicopter Maritime Strike Squadron 72 (HSM-72) after transitioning to the MH-60R. HSL-42 parented 10 detachments for operations aboard the US Navy's cruisers and destroyers.

MH-60R

Type: Anti-Surface and Anti-Submarine Warfare Helicopter
Dimensions: Length: 19.71m (64ft 8in); Rotor diameter: 16.36m (53ft 8in); Height: 5.23m (17ft 2in)
Weight: 10,433kg (23,000lb) maximum take-off
Powerplant: 2 × General Electric T700-GE-401C turboshaft engines 1410 kW (1890 shp) each
Maximum speed: 270km/h (168mph)
Range: 830km (520 miles)
Service ceiling: 3657m (12,000ft)
Crew: 3–4
Armament: 7.62mm (0.3in) M60 or M240 machine gun, or 12.7mm (0.5in) M2 heavy machine gun, AGM-114 Hellfire missiles, AGR-20 APKWS 70mm (2.75in) guided rockets, Mk46, Mk50 or Mk54 torpedoes

NAVAL HELICOPTERS

An SH-60B Seahawk from Helicopter Anti-Submarine Squadron Light Fifty-One (HSL-51), firing an AGM-119B Penguin Anti-Ship Missile. Penguin is a Norwegian designed and built missile with an infrared guidance system, a 130kg warhead, and a range of 55km (34mph).

NAVAL HELICOPTERS

Sikorsky SH-60F Oceanhawk
Helicopter Antisubmarine Squadron 3 (HS-3) operated the SH-60F from 1991. As part of Carrier Air Wing 8 (CVW-8), it deployed with the USS *Theodore Roosevelt* through the early 1990s, including during Operation Desert Storm. In 2009 the squadron converted to the MH-60S Knight Hawk and was redesignated Helicopter Sea Combat squadron 9 (HSC-9).

units to the MH-60R was effectively complete in 2015 when the SH-60B, SH-60H and HH-60H all ended their last operational deployments. The former was formerly retired and the latter two were relegated to second line units after their squadrons transitioned to newer models.

Jayhawks

While the SH-60B was intended to operate 60km (40 miles) or so ahead of a task group, the SH-60F was developed to provide close in ASW protection, replacing the SH-3H Sea King. Also known as the Oceanhawk, the SH-60F dispensed with the radar and MAD system while gaining an AN/AQS-13F dipping sonar and a simpler six sonobuoy launcher. Intended to operate from aircraft carriers, the SH-60F also did without the RAST. The SH-60F formed the basis of the HH-60H Combat SAR version, and the US Coast Guard's HH-60J and MH-60T medium-range SAR helicopters known as Jayhawks.

Achieving initial operating capability in 2006 the MH-60R Seahawk combines the capabilities of the SH-60B and SH-60F in one aircraft while also upgrading the systems. The original cockpit analogue gauges have been replaced by four NVG-compatible multifunction displays. The radar has been replaced with the AN/APS-153, which features an automatic periscope detection capability. An AN/AQS-22 dipping sonar replaces the early models, and from 2020 the CAE MAD-XR system has provided a MAD capability using a sensor mounted inside the tail rather than towed behind the aircraft. An electro-optical (EO) turret is mounted on a plinth in front of the nose and includes a laser rangefinder. Self-protection is provided by an AAR-47 missile approach and warning system combined with decoy dispensers and infrared jammers. As with the earlier Seahawks, offensive weaponry includes Mk 46 air-launched torpedoes and AGM-114 Hellfire air-to-surface missiles, while the pylons can also be used to carry external fuel tanks. The laser-guided APKWS 70mm (2.75in) rocket has since been integrated with the SH-60R along with the more modern Mk 54 torpedo. The transition of frontline

Knight Hawk

The HH-60H has been in part replaced by the MH-60S, the only naval variant to use the original Black Hawk tail wheel. Also known as the Knight Hawk, the MH-60S marries the fuselage of the UH-60L with the MH-60R's transmission, folding rotors and tail pylon, and avionics systems. Replacing the USN's Sea Knight, the S model is used for tasks such as logistics, combat SAR and mine detection using the Airborne Laser Mine Detection System (ALMDS). For offensive operations it can carry guns and rocket pods on stub wings with a FLIR and laser designator in a nose-mounted pod.

The Seahawk has been an export success with Japan alone ordering over 100 SH-60J aircraft based on the SH-60B. Several nations have replaced their earlier models with MH-60R while new customers are continuing to place orders, most recently Norway who ordered six in 2023 to replace the NH-90.

NAVAL HELICOPTERS

Sikorsky SH-3D Sea King

In 1957, in response to the increasing perceived threat from Soviet submarines, Sikorsky won the contract to develop a replacement anti-submarine warfare (ASW) helicopter for the US Navy (USN). This was to become the Sea King.

Most importantly the new aircraft utilized the recently developed T-58 turboshaft in place of the piston engines that had hitherto powered the USN's rotorcraft. This promised significant saving in space and weight while delivering similar power levels with less vibration. Where the Sikorsky HSS-1 had been powered by a Wright R-1820 Cyclone able to produce around 746kW (1000shp) while weighing 537kg (1184lb), the T-58 could produce a similar level of power while only weighing 177kg (391lb). The new engine was also much smaller, taking up only a sixth of the space.

True hunter-killer

The Sea King's predecessors had been forced to compromise on layout, placing the large piston engine in the nose with the drive shaft to the gearbox running through the cockpit, which sat over the forward end of the cabin. With the T-58, Sikorsky could design the aircraft around the needs of the crew and mission instead. The engines were placed on top of the fuselage in front of the main rotor gearbox, allowing the cockpit and cabin to be placed on the same level with easy access from one to the other. For the ASW mission an ASQ-10 low frequency dipping sonar was housed in the cabin and lowered through a hole in the fuselage floor to depths of 150m (500ft). Crucially, unlike its predecessors, the SH-3A was able to carry a sonar at the same time as its offensive weaponry of up to four lightweight torpedoes and operate as a true hunter-killer. To ease the pilot workload, an automatic stabilization system was fitted that was also capable of flying the helicopter into wind while hovering and maintaining position over the dipping sonar as it was deployed. A boat-shaped hull was chosen, along with sponsons mounted on stub wings, to improve the stability and survivability in the event the aircraft was forced to land on water.

Entering service with the USN in 1961 as the HSS-2, the Sea King was redesignated the SH-3A in 1962 under the new tri-service system. It was powered by the 937kW (1250shp) T58-GE-8B allowing a gross weight of over 8½ tonnes. Aside from the ASW mission, SH-3A were modified for the

Westland Sea King HU Mk5
Operated by 771 Naval Air Squadron (NAS) in a distinctive red-and-grey scheme, the HU Mk5 was used for SAR duties from Royal Naval Air Station (RNAS) Culdrose. ZA166 was converted from a HAS Mk5 airframe in 2001. Retired in 2017, it is believed to be one of three transferred to the Ukrainian Navy in 2023.

Sikorsky SH-3H
Type: Anti-Surface Warfare Helicopter
Dimensions: Length: 16.69m (54ft 9in); Rotor diameter: 18.9m (62ft); Height: 5.13m (16ft 10in)
Weight: 10,000kg (20,050lb) maximum take-off
Powerplant: 2 x General Electric T58-GE-10 1045kW (1400shp) each
Maximum speed: 267km/h (166mph)
Range: 1000km (621 miles)
Service ceiling: 4480m (14,700ft)
Crew: 4
Armament: 2 x Mk44/46 torpedoes, B57 nuclear depth charge.

NAVAL HELICOPTERS

combat SAR role in Vietnam as the HH-3A with titanium armour, external fuel tanks on the forward torpedo hardpoints and deletion of the ASW equipment. Nine were also converted to RH-3A standard with a second cargo door on the left-hand side of the fuselage and towing equipment for mine countermeasures equipment mounted under the tail boom. Although under powered for the role, they proved the concept that would be refined by the CH-53.

The SH-3D

Entering service in the mid-1960s, the next ASW version was the SH-3D powered by the 1050kW (1400shp) T58-GE-10. To absorb the extra power, the gearbox was strengthened allowing an increase in gross weight to 9.3 tonnes while an increased span braced horizontal stabilizer was taken from the US Air Force's CH-3C. The ASW equipment was significantly improved with the AN/AQS-13A sonar and AN/APN-182 Doppler navigation system, and on some aircraft lengthened sponsons were fitted to carry a towed magnetic anomaly detector (MAD). Although weapons carriage was reduced to two torpedoes, the SH-3D gained the ability to drop them from the hover as well as in forward flight. While only 72 SH-3D were built for the USN, in contrast to 245 SH-3A, it would be licence-built by Agusta, Mitsubishi and Westland Helicopters. It would also form the basis of the VH-3D presidential transport, known as Marine One, that remains in service to this day.

Sikorsky meanwhile developed the SH-3G for the USN as a transport aircraft, converting SH-3A and 3Ds

The German Navy received 23 Westland built Sea King Mk.41 between 1973 and 1975 for SAR operations. Based on the RN's HAS Mk1 but without sonar equipment, from 1986 a Ferranti Seaspray radar was added to the nose to guide Sea Skua anti-ship missiles.

NAVAL HELICOPTERS

by removal of the sonar equipment, reinforcing the cabin floor, and for VIP aircraft enlarging the cargo door window. Introduced in 1970 these were later replaced by the similarly modified UH-3H that remained in service into the 2000s, by which time they were based ashore in locations such as Bahrain to support US and Allied ships operating in the Gulf.

In 1971 Sikorsky began converting SH-3A, 3D and 3G, into the SH-3H with the first flight the following year. This further upgraded the type's ASW capabilities. The sonar was replaced by the AN/AQS-13B, a torpedo pre-setting system was added allowing different search profiles to be followed, the towed MAD was fitted to all aircraft, with 24 launch tubes for marine markers fitted in an enlarged port sponson, and sonobuoy launch tubes were fitted in the rear cabin. Additionally, a Litton LH-66 search radar was mounted in a retractable radome under the rear fuselage. This feature was to be short-lived, however, and was soon removed to save weight. One hundred and sixty-three aircraft were converted in total by Sikorsky while CASA of Spain undertook a similar upgrade to the Spanish Navy's SH-3Ds in the 1980s. While the SH-60F (see separate entry) started to replace the SH-3H from 1989, this took some time with seven squadrons using the Sea King in the ASW role during the 1991 Gulf War. The USN only finally retired the Sea King in 2006 by which time it had been in service for 45 years.

Westland's Sea Hawk

Westland Helicopters' licence production of the SH-3 in the UK led to a separate family of aircraft due to the generous nature of the agreement. The initial Sea King HAS Mk1 was

Sea King 89+62 is seen here at the 2018 ILA Berlin Air Show six years before the type's retirement by the German Navy. Six were donated to the Ukrainian Navy in August 2024 undergoing preparation work at HeliOperations in the UK prior to delivery.

NAVAL HELICOPTERS

Westland Sea King Mk.48
Used by the Belgian Air Force for search-and-rescue (SAR), RS02 was delivered to Koksijde in November 1976. During its service life it carried out 3309 SAR operations, saving 1757 lives. Retired in 2019 it is now owned by Historic Helicopters in the UK as part of a fleet of eight Sea Kings.

Westland Sea King Mk.48
Type: Anti-Surface Warfare Helicopter
Dimensions: Length: 16.69m (54ft 9in); Rotor diameter: 18.9m (62ft); Height: 5.13m (16ft 10in)
Weight: 10,000kg (20,050lb) maximum take-off
Powerplant: 2 x General Electric T58-GE-10 1045kW (1400shp) each
Maximum speed: 267km/h (166mph)
Range: 1000km (621 miles)
Service ceiling: 4480m (14,700ft)
Crew: 4
Armament: 2 x Mk44/46 torpedoes, B57 nuclear depth charge.

based on the SH-3D, however, unlike Sikorsky-built aircraft, they featured an Ecko AW391 surface search radar mounted above the aft fuselage in a thimble-shaped radome. Additionally, most of the systems were replaced with UK-developed alternatives. Rolls-Royce licence built T58 engines as the Gnome, although with an analogue fuel control computer developed by de Havilland Propellers. In line with Royal Navy (RN) doctrine, the UK's ASW Sea Kings were capable of operating autonomously with their own tactical co-ordinator – observer in RN parlance – to command the aircraft and other assets as required to fight the sub-surface threat. This contrasted with the USN's approach of treating the aircraft as an extension of the ship's sensors; consequently the observer had a tactical plotting display taking inputs from the radar and sonar.

The HAS Mk2 featured an improved six-bladed tail rotor in place of the original five-bladed unit. Engine, avionics and sensor upgrades were also incorporated with most of the 56 Mk1 airframes upgraded in addition to 21 new-build aircraft. From 1981 the HAS Mk5 was introduced, recognizable by its larger flat-topped radome for the MEL Super Searcher radar, while there were further upgrades to the ASW systems. Again, older airframes were upgraded to the Mk5 standard in addition to 30 new aircraft. The HAS Mk5 and Mk2 were operational during the 1982 Falklands conflict along with the HC Mk4 that had just entered service.

This was a dedicated tactical transport version able to carry up to 28 fully equipped troops. Unlike the ASW versions, it did not have sponsons on the stub wings and the undercarriage was fixed. One of these aircraft ended up in Chile after a one-way mission to insert special forces troops into Argentina.

Westland produced over 300 Sea Kings, many of which were exported to the Middle East, Australia, India and Pakistan. The latter two countries armed theirs with Sea Eagle and Exocet anti-ship missiles respectively, making them the most heavily armed variants produced. After the 1982 Falklands conflict, Westland also developed an AEW version of the Sea King for the Royal Navy, which carried a Searchwater radar in an inflatable dome that swung down below the aircraft once in flight. Initially designated the AEW Mk2, a significant upgrade at the start of the 21st century turned it into the ASaC Mk7, which remained in service until 2018.

Although the Sea King has now retired from service with the USN and RN, it remains in use with several countries. Most recently Ukraine has taken delivery of surplus Royal Navy aircraft for use in the SAR and utility roles. Meanwhile USN aircraft are being regenerated from the AMARC Bone Yard in Arizona for use with the State Department and other branches of the US Government or for sale to civilian and foreign operators.

Westland Sea King HAS Mk5
Dimensions: Length: 17.02m (55ft 10in); Rotor diameter: 18.9m (62ft); Height: 5.13m (16ft 10in)
Weight: 9707kg (21,400lb) maximum take-off
Powerplant: 2 x Rolls-Royce Gnome H.1400-1 turboshaft engine 1238kW (1660shp) each
Maximum speed: 267km/h (166mph)
Range: 1230km (764 miles)
Service ceiling: 4480m (14,700ft)
Crew: 4
Armament: 4 x Mk44/46 or Sting Ray torpedoes, depth charges, door mounted machine guns

Westland Sea King Mk42C
The Indian Navy operated 41 Westland-built Sea Kings along with six Sikorsky-built examples acquired in 2007. The Mk42C is a SAR version with a Bendix radar mounted on the nose. Replacement of the Sea King by the MH-60R Seahawk started in 2024.

AgustaWestland AW101 Merlin

Believing the Sea King could be obsolete in the anti-submarine warfare (ASW) role by 1987, the Royal Navy issued a requirement for a replacement in 1977.

This called for a helicopter capable of ASW, surface search and surveillance, vertical replenishment, search and rescue, and the Commando tactical transport role. Outline specifications were a cruise speed of 140kts with an endurance of four hours while 90km (56 miles) from its parent ship. Sensors for the ASW role included a 360° scanning radar, sonobouy dispenser and processing unit, and a dipping sonar, while four Sting Ray torpedoes would provide the striking power. For the Commando role carriage of at least 20, but preferably 34, troops in Arctic clothing or an internal load of 3200kg (7054lb) over a radius of action of at least 185km (115 miles) was required. While this called for an aircraft with an all-up mass of around 11 tonnes, it had to fit within the hangar of a frigate with maximum folded dimension of 15.8 by 5.5m (51.8 x 18ft) and a height of 5.2m (17ft), approximately the size of two double-decker buses parked next to each other.

Italian collaboration

With the cost of developing the aircraft on its own prohibitive, the UK sought an international collaborator and in 1979 a Memorandum of Understanding was signed with Italy. Both countries envisaged a requirement for a helicopter bigger than the proposed NH90. The resulting EH101 was of conventional configuration with a five-bladed main rotor and four-bladed

The Italian Navy operates eight SH-101A in the anti-submarine warfare role. Clearly visible in this shot is the opening for the dipping sonar on the underside of the fuselage in line with the forward edge of the undercarriage sponsons.

NAVAL HELICOPTERS

Cockpit
Although designed to be operated by a single pilot, the Merlin is almost always operated with two. In ASW versions the observer and aircrewman's consoles are in the forward cabin adjacent to the cockpit, with the dipping sonar aft of them.

Rotors
The main rotor has composite blades with paddle tips to reduce the onset of retreating blade stall. This occurs when the forward speed of the aircraft reduces the airflow over the rearwards travelling blade below its stall speed.

Radar
AW101 can be fitted with 360° radar under the forward fuselage; on the Royal Navy's Merlin this is the Blue Kestrel 5000 surveillance radar. Some operators have instead chosen a forward-facing radar mounted on the nose.

tail rotor. The name was changed to AW101 when Agusta and Westland merged in July 2000. The main rotors are based on the British Experimental Rotor Programme (BERP) with enlarged spade-like tips that improve the operational flight envelope. Power is provided by three engines either, RTM 322 or T700s depending on the customer, with the third engine mounted behind the main rotor gearbox and its inlet just to the left of the rotor mast. The glass cockpit originally featured eight multifunction displays, although upgrades to the Royal Navy's fleet have seen this rationalized to five larger units on the main panel with two further touchscreens on the centre console replacing an array of switches. The airframe is a mix of aluminium alloy and composites, and the tail is all composite while the fuselage features a metal frame covered in alloy-skinned composite panels. Despite being a large helicopter, the requirement to land on a frigate in up to a sea state 6 means the Merlin is also highly manoeuvrable.

Although the first pre-production aircraft flew in October 1987, development was not straightforward. The crash of two pre-production examples in the early 1990s led to redesign work that delayed the start of production. It was not until 1997 that the first aircraft were delivered to the Royal Navy's 700M trials squadron. The first operational squadron, 814 NAS finally formed in June 2000; by this time the original RN requirement for 66 Merlin HM Mk1 had been reduced to 44 due to the end of the Cold War. The Italian Navy's aircraft entered service during 2001, the same year the RAF began operation of the HC Mk3 troop-carrying version that initially replaced the Wessex.

NAVAL HELICOPTERS

EH101 PP5

The fifth pre-production EH101, and fourth built by Westland, PP5 was used to develop the avionics fit for the Royal Navy's Merlin HM Mk1.

Merlin HM Mk2

Type: ASW, SAR and Utility Helicopter
Dimensions: Length: 22.8m (74ft 10in); Rotor diameter: 18.59m (61ft); Height: 6.6m (21ft 10in)
Weight: 14,600kg (32,187lb) maximum take-off
Powerplant: 3 × Rolls-Royce Turbomeca RTM322-01 turboshaft engines, 1566 kW (2100shp) each
Maximum speed: 309km/h (192mph)
Range: 1390km (863 miles)
Service ceiling: 4575m (15,000ft)
Crew: 4
Armament: 1 x 12.7mm (0.5in) M3M heavy machine gun, 4 x Sting Ray homing torpedoes, Mk11 depth charge

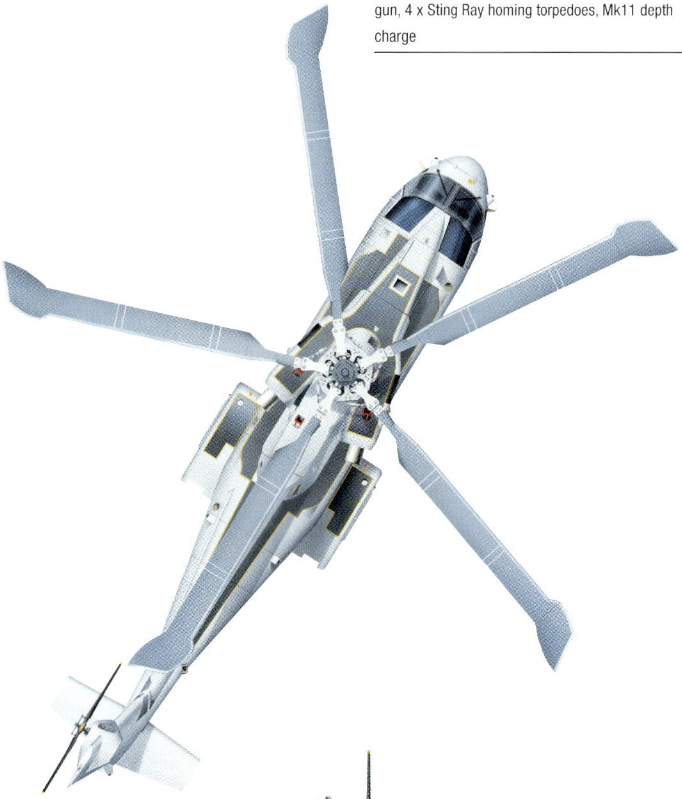

Deployment

Royal Navy and RAF Merlins saw extensive use in the Persian Gulf, Iraq and Afghanistan with a range of capabilities being added for those theatres. These included an AN/AAR-57 missile warning system, AN/ALQ-157 infrared jammers, and AN/ALE-47 chaff and flare dispensers. Wescam MX-15 electro-optical turrets were also acquired for use by deployed Merlin HM Mk1s. When necessary, the ASW version can be reconfigured as transports by removal of the sonar equipment, which frees up space for additional passenger seating

A Royal Navy Merlin HC Mk4 departs the USS *Bataan* in the Eastern Mediterranean in early 2024. The HC Mk4 has a rear loading ramp to access the cabin, while the upgrade from Mk3 standard included full marinization and an avionics upgrade.

and stretchers. In 2014 the RAF's Merlin Mk 3 transport aircraft began being transferred to the Royal Navy's Commando Helicopter Force and have been upgraded to Mk4 standard. This adds blade folding and a common cockpit with the upgraded Mk2 ASW aircraft for better ship integration.

Similar to the RN, the Italian Navy operates an ASW model, the SH-101A, and an amphibious support model, the MH-101A. The Italian Air Force meanwhile has a fleet of 12 HH-101A Caesar aircraft for Combat SAR.

Although the primary goal of the EH101 programme was the replacement of ASW helicopters, the requirement to succeed the Sea King in its other roles has led to a capable transport and SAR aircraft. Unlike the Black Hawk family, the cabin has standing headroom, and it can carry 30 troops, a 3-tonne internal payload, or a 5.5 tonne underslung load. In fact, the majority of foreign sales have been for dedicated SAR aircraft with Canada, Norway and Portugal, buying them solely for this purpose while Denmark has a mixed SAR and tactical transport fleet. Poland and Japan have acquired naval Merlins, the former for the ASW role while Japan chose them for the airborne mine countermeasure (AMCM) role replacing the MH-53E. For this they are equipped with an AQS-24A mine hunting system, AN/AES-1 airborne mine detection system and the ability to tow a Mk-104 acoustic sweep.

Glossary

AAFSS	Advanced Aerial Fire Support System		FADEC	Full Authority Digital Engine Control
AAH	Advanced Attack Helicopter		FLIR	Forward Looking Infra-Red
AESA	Active Electronically Scanned Array		FOB	Forward Operating Base
AEW	Airborne Early Warning		HELRAS	Helicopter Long Range Active Sonar
AHIP	Army Helicopter Improvement Programme		HIRF	High Intensity Radiated Fields
			HIRSS	Hover Infra-Red Suppression System
ALMDS	Airborne Laser Mine Detection System		HIRTA	High Intensity Radio Transmission Area
AMCM	Airborne Mine Counter Measures			
APKWS	Advanced Precision Kill Weapon System		HOT	High Subsonic, Optical, Remote-Guided, Tube-Launched
ASaC	Airborne Surveillance and Control		ILS	Instrument Landing System
ASUW	Anti-Surface Warfare		LMM	Lightweight Multirole Missile
ASW	Anti-Submarine Warfare		MAD	Magnetic Anomaly Detector
BERP	British Experimental Rotor Programme		MMS	Mast Mounted Sight
			NFH	NATO Frigate Helicopter
CEP	Circular Error of Probability		NVG/NVD	Night Vision Google/Night Vision Device
CMRB	Composite Main Rotor Blade			
CTS	Central Tactical System		PNVS	Pilot's Night Vision System
DAFCS	Digital Automatic Flight Control System		RAST	Recovery Assist Secure and Traversing
DAP	Direct Access Penetrator		SAM	Surface to Air Missile
DAS	Defensive Aids Suite		SAR	Search and Rescue
DIRCM	Directed Infra-Red Counter Measure		SOAR	Special Operations Aviation Regiment
DSP	Digital Signal Processor		TADS	Target Acquisition and Designation Sight
EAPS	Engine Air Particle Separators			
ECM	Electronic Counter Measures		TOW	Tube-launched, Optically tracked, Wire-guided
EO	Electro Optical		TTH	Tactical Transport Helicopter
ESM	Electronic Support Measures		UHF	Ultra-High Frequency
ESSS	External Stores Support System		VTOL	Vertical Take-Off and Landing
EW	Electronic Warfare			

Index

Index note: page numbers in *italics* refer to images or their captions.

A

Abu Dhabi 68
ACH-47A 88–9
Aérospatiale
    SA330 Puma 68, 72–3
    SA341/342 Gazelle 56–8
Afghanistan 13, 26–7, 45, 48, 52, 53, 63, 78, 86, 121
Agusta-Bell AB-204 *44–5*
AgustaWestland
    AW101 Merlin 13, 119–22
    AW139 76
    AW159 Wildcat 104–6
AH-1 Cobra 7, 10, 20, 40, 45, 52, 89
AH-1 Super Cobra 22–4
AH-1J Sea Cobra 22
AH-1T 22–3
AH-1Z Viper 25
AH-6 Little Bird 20–1
AH-56A Cheyenne 10
AH-64 Apache *8, 10,* 10–15, *13, 14–15,* 23, 40, 52
    AH-64A *11,* 12
    AH-64D *10,* 12–13, *13, 14–15*
    AH-64E *8, 13,* 13–14, 16
Airbus H145M 46
Airbus Helicopters EC665 16
Algeria 32, 82
Apache *8,* 10–15, 16, 23, 40, 52
Argentina 73, 107, 118
Army Helicopter Improvement Programme (AHIP) 50–1

AS332 Super Puma 68–9, *72*
AS350 Squirrel 61
AS365 Dauphin II 54–5
AS532 Cougar 68–9
AS565 Panther 54–5
attack helicopters 8–35
    AH-6/MH-6 Little Bird 20–1
    AH-64 Apache *8, 10,* 10–15, *13, 14–15*
    Bell AH-1 Super Cobra 22–4
    Bell AH-1Z Viper 25
    Changhe Z-10 33
    Eurocopter Tiger 16, 33
    HAL Prachand 34
    Harbin Z-19 35
    Kamov Ka-50/Ka-52 17–19
    Mil Mi-24 26–8
    Mil Mi-28N 30–2
    Mil Mi-35M 28–9
Australia 16, *43,* 47–8
AW101 Merlin 13, 119–22
AW139 76
AW159 Wildcat 104–6

B

Bahrain 25, 49, 59
Beirut 23
Belarus *80*
Bell
    412 59–60, 62
    AH-1 Super Cobra 22–4
    AH-1Z Viper 25
    CH-146 Griffon 53
    OH-58 Kiowa 50–2
    UH-1 Huey *6, 6,* 7, 43–5, 59, 77
    UH-1Y Venom 62–3

Bell-Boeing V-22 Osprey 85–7
Bell Griffin HAR Mk2 *60*
Black Hawk *36,* 38–42, 46, 48, 111
Bo 105 49
Boeing
    AH-6i *21*
    AH-64 Apache *8, 10,* 10–15, *13, 14–15,* 23, 40, 52
    CH-47 Chinook 6, 7, 40, *70,* 88–93
    MH-6M *20*
Bölkow Bo 105 49
Bosnia 53
Brazil 55, 68, 69, 107
Bristol Sycamore 6
Britain 57–8, 73, 86

C

C-130 Hercules 21, 38
Canada 53, 59, 122
Caracal 68–9
CH-47 Chinook 6, 7, 40, *70,* 88–93
    CH-47B 89–90
    CH-47C 90
    CH-47D 90–2
    CH-47F 91, 93
CH-53 Sea Stallion 100–3
    CH-53D 100–2
    CH-53E Super Stallion *100–1,* 102–3
    CH-53K King Stallion *100,* 103
CH-146 Griffon 53
Changhe
    Z-8 74–5

Z-10 33
Z-11WB 61
Z-18 74
Cheyenne 10
Chile 60, 68, 73
China 33, 35, 61, 65–7, 72, 74–5, 81–2
Chinook see CH-47 Chinook
CMV-22B 86–7
Cobra 7, 10, 20, 40, 45, 52, 89
Cougar 68–9
CV-22B 86–7
Czech Republic 25, 62–3

D
Dauphin II 54–6
Denmark 107, 108, 122
Dhruv 34
Dolphin 54
Dominican Republic 55

E
EC135 49
EC155 55
Egypt 17, 19, 58, 80
EH-60A Electronic Warfare 41
EH101 119–20, 121
electronic warfare (EW) aircraft 41, 78
Eurocopter
  AS350 Squirrel 61
  AS365 Dauphin II 54–5
  AS532 Cougar 68–9
  EC135 49
  EC155 55
  Tiger 16, 33

F
F-117 stealth fighter 12
Falklands conflict 73, 108, 118
Fiery Thunderbolt 33
France 55, 56–8, 68, 72–3, 107

G
Gazelle 56–8
Germany 6, 16, 46, 47–8, 49, 69, 107, 116
Greece 52, 68
Grenada 21, 22–3, 40
Griffon 53
Guatemala 59
Gulf of Aden 55
Gulf War 12, 23, 40, 52, 57, 121

H
H145M 46
H160 55
H225M Caracal 68–9
Haiti 23, 53
HAL Prachand 34
Harbin
  Z-9 65
  Z-19 35
  Z-20 66–7
  Z-21 67
Havoc-B 19, 30–2
HC Mk2 90, 92
heavy lifters see transporters and heavy lifters
Helix 96–8
Hercules 21, 38
HH-65A Dolphin 54
Hind 26–8, 58
Hip 7, 26–7, 77–82
Hoodlum 64

Huey see UH-1 Huey
Hughes OH-6 52

I
India 34, 97, 118
Indonesia 49, 59, 73, 81
Iran-Iraq war 21, 58
Iraq 13, 52, 58, 63, 86 see also Gulf War
Iraq War 21, 24
Iroquois see UH-1 Huey
Israel 11, 13, 55, 64
Italy 44–5, 47, 47–8, 59, 76, 119–20, 122

J
Japan 86, 113, 122
Jayhawk 113
Jordan 21, 83

K
Kamov
  Ka-27 96–8
  Ka-28 97–8
  Ka-29TB 97
  Ka-31 99
  Ka-32T 98
  Ka-50/Ka-52 17–19
  Ka-226 64
Kawasaki CH-47J Chinook 92–3
King Stallion 100, 103
Kiowa 50–2
Knight Hawk 113
Kuwait 58, 108

L
Lakota 46
Laos 98

INDEX

Lebanon 23, 58
Leonardo Helicopters 76
Libya 16, 90
Little Bird 20–1
Lynx 58, 105, 107–10

M

Macedonia 78–9
Mali 16, 86
Marine One 115
McDonnell Douglas AH-64 Apache 11
MD 530F 21
Merlin 13, 119–22
Mexico 55
MH-6 Little Bird 20–1
MH-47 Chinook 88–9, 92–3
MH-53E Sea Dragon 102–3
MH-53J 12, 100–2
MH-60R Seahawk 111, 113
MH-60S Knight Hawk 113
MH-60T Jayhawk 113
Mil Mi-8 Hip 7, 26–7, 77–82
Mil Mi-17 77, 81
Mil Mi-24 26–8, 58
Mil Mi-25D 26, 28–9
Mil Mi-26 83
Mil Mi-28N 19, 30–2
Mil Mi-35M 7, 28–9
Mil Mi-35P 29
Mil Mi-38 84
Mil Mi-171 82

N

naval helicopters 7, 94–122
    AgustaWestland AW101 Merlin 13, 119–22
    AgustaWestland AW159 Wildcat 104–6
    CH-53 Sea Stallion 100–3
    Kamov Ka-27 96–8
    Kamov Ka-31 99
    Sikorsky SH-3D Sea King 114–18
    Sikorsky SH-60 Seahawk 111–13
    Westland Lynx 58, 105, 107–10
The Netherlands 47–8, 69, *110*
New Zealand 48
NHIndustries NH-90 47–8
Nigeria 34
Northern Ireland 58
Norway 48, *60,* 107, 113, 122

O

Oceanhawk *94, 112–13,* 113
OH-6 52
OH-58 Kiowa 50–2
OH-58D 50–2
Oryx 73
Osprey 85–7

P

Pakistan 33, 65
Panama 12, 21, 40
Panther 54–5
Persian Gulf 23, 51–2, 121
Poland 76, 122
Portugal 73, 110, 122
Prachand 34
Puma 68, 72–3

Q

Quick Fix 41

R

R-4 6
radar 7
reconnaissance and utility helicopters
    Aérospatiale SA-342 Gazelle 56–8
    Airbus H145M 46
    Bell 412 59–60, 62
    Bell CH-146 Griffon 53
    Bell OH-58 Kiowa 50–2
    Bell UH-1 Huey 6, *6,* 7, 43–5, 59, 77
    Bell UH-1Y Venom 62–3
    Bölkow Bo 105 49
    Changhe Z-11WB 61
    Eurocopter AS365 Dauphin II 54–5
    Eurocopter AS532 Cougar 68–9
    Harbin Z-9 65
    Harbin Z-20 66–7
    Kamov Ka-226 64
    NHIndustries NH-90 47–8
    UH-60 Black Hawk 36, 38–42, 46, 48, 111
Romania 73
Rooivalk 73
Royal Navy 104–10, *114,* 117–22
Russia 7, 18–19, 28–9, 30–2, 64, 76, 77–84, 98–9
    see also Soviet Union

S

S-61 Sea King 114–18
SA330 Puma 68, 72–3
SA341/342 Gazelle 56–8, *58*

INDEX

Saudi Arabia *21,* 55, 68, 69
Sea Cobra 22
Sea Dragon 102–3
Sea King 114–18
Sea Stallion *see* CH-53 Sea Stallion
Seahawk *see* SH-60 Seahawk
SH-3D Sea King 114–18
SH-60 Seahawk 111–13
SH-60B Seahawk 40
SH-60F Oceanhawk *94, 112–13*
Sikorsky
  CH-53 Sea Stallion 100–3
  R-4 6
  SH-3D Sea King 114–18
  SH-60 Seahawk 111–13
  UH-60 Black Hawk *36,* 38–42, 46, 66, 111
South Africa 73, 110
South Korea 49, 55
South Sudan 86
Soviet Union 17–18, 26–8, 77, 99
Spain 68, 116
Squirrel 61
Sud Aviation SA330 Puma 68, 72–3
Suez crisis, 1956 6
Super Cobra 22–4
Super Lynx 110
Super Puma 68–9, *72*
Super Stallion *100–1,* 102–3
Sweden 48
Syria 31, 58, 64, 79, 99

T

Tanker War, Persian Gulf 23
Thailand 21, 46
Tiger 16, 33
transporters and heavy lifters 71–93
  Aérospatiale SA330 Puma 68, 72–3
  AgustaWestland AW139 76
  Bell-Boeing V-22 Osprey 85–7
  Boeing CH-47 Chinook 6, 7, 40, *70,* 88–93
  Changhe Z-8 74–5
  Mil Mi-8 7, 26–7, 77–82
  Mil Mi-17 77, 81
  Mil Mi-26 83
  Mil Mi-38 84
  Mil Mi-171 82
Turkey 68

U

UH-1 Huey 6, *6,* 7, 43–5, 77
  UH-1N Twin Huey 22, 45, 59, 62
  UH-1Y 25, 38
  UH-1Y Venom 62–3
UH-60 Black Hawk *36,* 38–42, 46, 48, 111
UH-72 Lakota 46
Ukraine 18–19, 28–9, 31–2, 64, 79, 97, *114,* 118
United Kingdom 57–8, 73, 86
  Royal Navy 104–10, *114,* 117–22
US Air Force 42, 45, 76, 86
US Army 10–13, 20–1, 38–42, 45, 46, *78*
US Coast Guard 54, 113
US Marine Corps *6,* 22–4, 25, 63, 86, 100–3
US Navy 111–13

V

V-22 Osprey 85–7
V-114 88
Venezuela 59
Venom 62–3
Vertol V-114 88
VH-3D Marine One 115
Vietnam *98*
Vietnam War 6–7, 20, 39, 43–5, 50, 88, 115
Viper 25

W

Westland *see also* AgustaWestland
  Lynx 58, 105, 107–10
  SA-341B Gazelle AH.1 *56–7, 57*
  Sea King *114,* 116–17
  WAH-64D Apache *12,* 13
Whirlwind 6
Wildcat 104–6
world record, long distance flight 20

Y

Yugoslavia 23, 58

Z

Z-8 74–5
Z-9 65
Z-10 33
Z-11WB 61
Z-18 74
Z-19 35
Z-20 66–7
Z-21 67

Picture Credits

Photographs:

AgustaWestland: 105 bottom

Alamy: 50 (Stocktrek Images), 52 (Pieter Stroobach)

Dreamstime: 19 & 32 (Artyom Anikeev), 69 (VanderWolfImages), 84 (Milkagenka), 103 (Walter Arce), 110 (Ryan Fletcher), 115 & 116 (VanderWolfImages)

Getty Images: 75 (VCG)

Shutterstock: 7 (Fasttailwind)

UK MOD Crown copyright 2024: 106 (CPL Katrina Knox)

U.S. Air Force: 27 (MSGT Steven Turner), 41 (TSGT Jeffrey Allen), 58 (SA Blake Wiles), 76 (SA Matthew-John Braman), 78 (SRA Jerry Morrison), 80 (SSGT Cherie A Thurlby), 87 (SSGT Christopher Callaway), 89 (MSGT Lance Cheung), 90 (SA Steven R Doty)

U.S. Air National Guard: 22 (MSGT Matt Hecht), 42 (PFC Paul Berzinas)

U.S. Army: 8 (CWO3 Mark Leung), 36 (SGT Steven Galimore), 70 (SSGT Michael Bracken), 91 (SPC Patrick Tharpe)

U.S. Coast Guard: 104 (PO2 Joseph Perrone)

U.S. Marine Corps: 6 (LCPL Nathan L Barnes), 24 (SGT William Waterstreet), 44 (LCPL Cory A Tepfenhart), 101 (SSGT Artur Shvartsberg), 102 (LCPL Molly Hampton), 122 (CPL Michele Clarke)

U.S. Navy: 92 (PM2 Timothy Smith), 94 (PH2 Milosz Reterski), 97 (MC2 Jason R Zalasky), 105 top (MCS3 Maxwell Orlosky), 112 (PH2 Lisa Aman), 119 (MCS2 Keith Nowak)

Artworks:

Edward Jackson/artbyedo: 66–67

Amber Books: 10–11, 14–15, 16, 23, 27, 30-31, 39, 48 (top), 51, 56 (top), 112–113, 120–121

Rolando Ugolini: 5, 12–13, 24–25, 38, 40–45, 47, 48 (lower), 49, 52–55, 56–57, 59–60, 62–63, 65, 68–69, 72–73, 88–93, 107–111, 114–118

Teasel Studios: 17–21, 26, 28–29, 33–35, 46, 61, 64, 74–87, 96–101